Regional Conference on Nuclear Physics (RCNP 2016)

EPJ Web of Conferences Volume 156 (2017)

Johor Bahru, Malaysia
15 - 16 December 2016

Editors:

Koh Meng Hock
Ganesan Krishnan
Faizani Mohd Noor

ISBN: 978-1-5108-5010-1

Printed from e-media with permission by:

Curran Associates, Inc.
57 Morehouse Lane
Red Hook, NY 12571

CURRAN ASSOCIATES INC. proceedings.com

Some format issues inherent in the e-media version may also appear in this print version.

Printed by Curran Associates, Inc. (2017)

For additional information, please contact EDP Sciences – Web of Conferences at the address below.

EDP Sciences – Web of Conferences
17, Avenue du Hoggar
Parc d'Activité de Courtabœuf
BP 112
F-91944 Les Ulis Cedex A
France

Phone: +33 (0) 1 69 18 75 75
Fax: +33 (0) 1 69 28 84 91

contact-edps@webofconferences.org

Additional copies of this publication are available from:

Curran Associates, Inc.
57 Morehouse Lane
Red Hook, NY 12571 USA
Phone: 845-758-0400
Fax: 845-758-2633
Email: curran@proceedings.com
Web: www.proceedings.com

TABLE OF CONTENTS

Introduction

The Regional Conference on Nuclear Physics (RCNP) 2016 has been organized for the first time immediately after the 3rd edition of the Winter School. This conference was co-organized by Universiti Teknologi Malaysia, Universiti Kebangsaan Malaysia and Centre Etudes Nucléaires de Bordeaux Gradignan, France. The aim of this conference is to provide a platform for local academicians, researchers and postgraduates to showcase their nuclear physics related researches. The topics covered in the conference are mainly related to applied nuclear physics with some contributions from fundamental studies in theoretical and experimental works.

edp sciences

EPJ Web of Conferences

Statement of Peer review

In submitting conference proceedings to EPJ Web of Conferences, I certify to the Publisher that I adhere to the **Policy on Publishing Integrity** of the journal in order to safeguard good scientific practice in publishing.

1. All articles have been subjected to peer review administered by the proceedings editors.
2. Reviews have been conducted by expert referees, who have been requested to provide unbiased and constructive comments aimed, whenever possible, at improving the work.
3. Proceedings editors have taken all reasonable steps to ensure the quality of the materials they publish and their decision to accept or reject a paper for publication has been based only on the merits of the work and the relevance to the journal.

Title, date and place of the conference

Title : Regional Conference on Nuclear Physics 2016
Date : 15th December 2016 to 16th December 2016
Place : T05, Faculty of Science, Universiti Teknologi Malaysia

Proceedings editor(s):

Koh Meng Hock
Ganesan Krishnan
Faizani Mohd Noor

Date and editor's signature

[illegible]/9/2017   25/07/2017   20/09/2017

Thermoluminescence Properties of Aluminium Oxide doped Strontium, Lithium and Germanium prepared by Combustion Synthesis method

Nurul Syazlin Binti Saharin*, N.E Ahmad, H.A Tajuddin, Abdul Rahman Tamuri

Department of Physics, Faculty of Science, Universiti Teknologi Malaysia, 81310 UTM Johor Bahru, Johor, Malaysia.

Abstract. In this work, Al_2O_3 doped with Sr, Li and Ge was prepared by combustion synthesis techniques for thermoluminescent (TL) ionizing radiation dosimetry applications. Dopants sample concentration were varied from 0.1 mol% to 5 mol% in order to study the effect of the concentration on the TL response. The optimum concentration was found to be 0.3 mol% for Sr and Li and 3.0 mol% for Ge. Based from the comparison of this doped material, the highest sensitivity was found for Ge doped and linear response for doses studied range of 10-80 Gy. X-Ray diffraction (XRD) patterns of the samples were recorded to confirm the formation of the sample. The sharp peaks present in the XRD confirms the sample is crystalline. The TL response of these samples for Co -60 gamma radiation were evaluated. It was observed a TL glow peak around 175°C for doped sample which is suitable for radiation dosimetry.

Keywords: Combustion synthesis, Aluminium oxide, Thermoluminescent dosimeter,

1 Introduction

One of the materials that can be considered for thermoluminescence TLD is aluminium oxide Al_2O_3, as it offers an excellent mechanical property and good mechanical stability [1]. Since the 50's, Al_2O_3 has been used as a radiation detection and dosimetric material using thermoluminescence technique. Onother advantage properties of Al_2O_3, as linearity in wide dose range, easy handling and more reasonable price [2]. Other than that, Al_2O_3 attracted scientific interest because of its high optical, chemical and thermal stability under irradiation. The challenges were to make it sensitive to radiation [3]. Akselrod *et al.*, (2003) reported that Al_2O_3 single crystals (sapphire, or α- Al_2O_3) as a material for volumetric optical data storage has several fundamental advantages. Since aluminium oxide shows an effective number close to that of human tissue (Z_{eff} = 11.28), it becomes a better choice to be used in medical and environmental dosimetry [4].

The conventional fabrication process of these dosimeters utilizes Czochralsky or Vernuil crystal growth technique, which involves the use of high temperatures (> 2000°C) and highly reducing atmosphere [5,6,7]. With the development of newer technologies, there are several other techniques to synthesize the material with low cost and easy to handle [7] such as sol-gel technique [8] on beam implantation [9], solvent evaporation [10] and combustion synthesis (CS) [5.11,12]. Among wet chemical methods, CS is an effective, and low-cost method to produce various industrially useful materials [13,14]. The solution preparation for combustion synthesis is quick and easy [15]. Furthermore, it would produce a highly pure and homogeeous powder. There are few studies done on this method, such as [16] where un-doped and Al (2-10 mol%) doped ZrO_2 nano powder were synthesized using solution combustion method having glycine as a fuel in a pre-heated sample of around 500°C. However, the study only concerned on the optical properties of the samples and is not focused on the thermoluminescence properties. Recently, to have a better performance in the dosimetric purposes, Al_2O_3 is always doped with impurities that induce many different types of trapping centers at which charged particles would produce ionizing radiation [15]. Since then, there are a lot of efforts have been directed towards the improvements on its sensitivity via various dopants such as Si, Ti, [17], Mg and Y [18] Cr and Ni [19]. Even though studies of dopant in Al_2O_3 has been done, search for new detector material for radiation monitoring as personal dosimeter is one of the important task in TLD research.

*Corresponding author: syazlin3@gmail.com

From previous study Lithium tetraborate, $Li_2B_4O_7$ (LTB) is a good tissue-equivalent material for thermoluminescent dosimetry having effective atomic number Z_{eff}= 7.25 which closely matches with that of human tissue [20], It has also been reported that the doped LTB materials are promising quantum electronics materials [21]. LTB is a versatile insulating material with large band gap [22]. In brachytherapy, the thermoluminescence dosimetry (TLD) using LiF crystals, is an experimental method that has been extensively applied with relative source.

The strontium aluminates have attracted intense research around the world, as they offer excellent properties such as high quantum efficiency, long persistence of phosphorescence and good stability. The long-lasting phosphors (LLP) oxide materials have been developed to replace the conventional sulfide after glow materials because of their improved luminescent properties such as high initial brightness, long lasting time, suitable emission colour and satisfactory chemical stability [23]. However, thermolumiescence properties of Ge doped Al_2O_3 prepared by CS method have not been reported so far. Recently, interest of Ge as dopants for dosimetric purpose has increased. Based on previous study, Ge doped optical fibre as a thermoluminescence dosimeter in radiotherapy is a promising material because it gives good response an fullfill the criteria of thermoluminescence properties [24]. Due to limitation study of CS technique especially for Al_2O_3 doped with Li, Sr and Ge dopants, for this present investigation, we report preliminary studies on the possibility of growing Al_2O_3 with different dopants and characterizing them for thermoluminescnet studies. We tried to explore different type of dopants such as Li, Sr and Ge doped for aluminium oxide to choose the best one for future study in thermoluminescence properties prepared by CS method.

2 Materials and Methods

2.1 Sample Preparation

For preparing powder samples, stoichiometric amount of aluminium nitrate, ($Al(NO_3)_3$) urea, ($CO(NH_2)_2$) (Sigma Aldrich reagent grade, 99.9% pure) and dopant Sr, Li and Ge were weighed using Shimadzu ATX 224 single Pan Analytical balance. The starting materials were a mixture of 15 g of aluminium nitrate, ($Al(NO_3)_3$), 6 g of urea, ($CO(NH_2)_2$) and different weight of dopant concentration from 0.1% to 5%. Then, 20 mL distilled water was added to the sample and magnetic stirrer was used to prepare a homogeneous solution stirred for 15 minutes, for completely dissolved before transferring to alumina crucible. An alumina crucible is used as a container within which chemical reaction take place. After the solution is transferred into the crucible, it is placed into a muffle-furnace at 500°C. Within 10 minutes, the reaction starts giving yellowish flames. The reaction was continuing for next few seconds and as it over, crucible was taken out from the furnace and kept in an open to allow cooling. Upon cooling, white foamy product formed which was then crushed using agate pestle mortar to get material in the powder form.

2.2 Measuring instruments

The XRD is used to confirm the crystalline nature of the as synthesized Al_2O_3 and to identify the presence of any impurity phases of different compositions from dopant material using Siemens Diffractometer D5000. The XRD machine is equipped with diffraction software with CuKα radiation operating at 40 kV, 30 mA with a scanning step length of 0.05° for 2 second counting per step and scanning angle from 10° to 90° with Bragg-Brentano geometry at room temperature. Annealing process was done to removes all residual of TL signals to re-establish TL sensitivity and eliminates unstable low-temperatures glow peak before sample is exposed. Al_2O_3 powders were annealed at 400°C for 1hour followed by cooling process at 80°C for 16 hours [25]. To evaluate the TL response, samples with different dopant at different concentration were irradiated at 50 Gy to Cobalt-60 gamma Cell Model 220 Excel at Universiti Kebangsaan Malaysia, Bangi, Malaysia. The TL response glow curve was measured using a TLD reader Harshaw Model 3500. The TL measurement was carried out at 300°C with heating rate of 3°C s^{-1}. All reading was taken under N_2 gas flow, to ensure good thermal contact between the sample and the heater-planchette, as is customary.

3 Results and Discussion

3.1 X-Ray diffraction (XRD) analysis

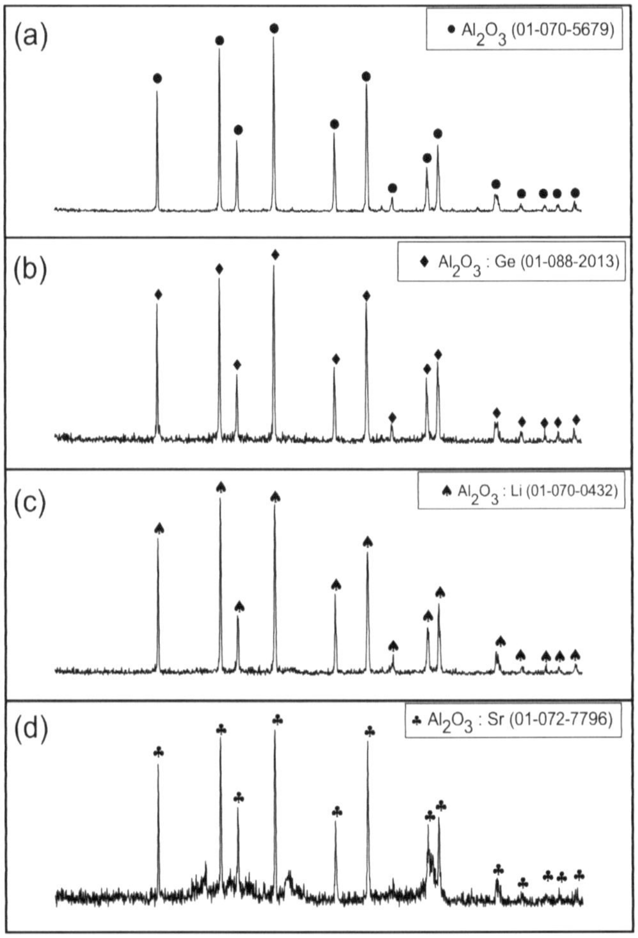

Fig 1. XRD pattern of (a) un-doped Al_2O_3, (b) Al_2O_3 doped with 3 mol% of Ge samples, (c) Al_2O_3 doped with 0.3 mol% of Li samples, (b) Al_2O_3 doped with 0.3 mol% of Sr samples

The XRD results prove that the prepared samples were crystalline and the respective peaks with their ICDD reference number is shown in Figure. 1 in this work. Changes of dopants do not contribute to any phase change of Al_2O_3. This finding is also supported by the previous work in [26]. They prepared Al_2O_3 by solution combustion method but using glycine as fuel.

3.2 Determination for the optimum strontium dopant concentration

Figure 2(a) represents the TL intensity and corresponding to standard deviation of Al_2O_3 at different Sr concentration. Figure 2(b) represents the glow curve of the samples at 50 Gy dose. From the results, the TL intensity of the sample was found to vary with decrease Sr concentration from 0.2 mol%. The intensity increases back from 0.3 mol% 531516.04 (a.u) and it is found to be optimum at this concentration. Based from the TL glow curve as in Figure 2 (b), the general nature of TL curves is similar in different cases: each curve peaks at temperature around 175°C. It is observed that varying concentration of Sr does not affect the peak position much. However, the emission intensity is found to increase with increasing Sr concentration and for x=0.3 mol%, the intensity is quenched and decrease for further concentration. The reason for such quenching may be the increase in probability of non-radiative transitions of the luminescent molecules from the excited state to the ground state in comparison to the probability of radiative transition. No such work for Al_2O_3: Sr has been reported to the best of our knowledge. From previous study this material has been considered to be effective luminescent hosts due to their low temperature synthesis, excellent chemical, physical stability and environmental stability [27]. As a conclusion, it is found that the optimum concentration value of Sr dopants is at 0.3 mol% (535116.04) with very low standard error.

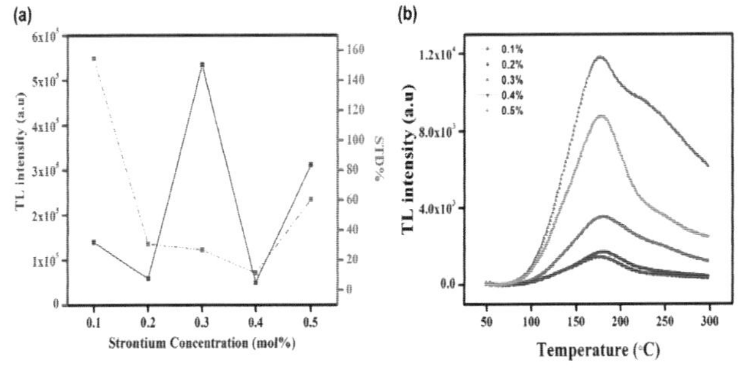

Fig 2. (a) TL intensity and correspond standard deviation as a function of different Sr concentration mole (0.1 to 0.5% mol concentration) (b) Glow curve recorded at different concentration mole (0.1 to 0.5%) of Sr dopant at a constant exposed to dose 50Gy

3.3 Determination for the optimum lithium dopant concentration

Figure 3(a) represents the TL intensity and corresponding to standard deviation of Al_2O_3 at different Li concentration. Figure3(b) represents the glow curve of the samples at 50 Gy dose. It can be seen from the Figure 3 (a) the TL intensity increase up to 0.3 mol% for Li concentration 126515.14 (a.u). Then the intensity decreases with further increase of Li concentration. This may be due beyond 0.3 mol% concentration, quenching is bound to occur hence the optimum concentration for Li only suite for this concentration. As conclusion, the highest TL intensity were observed for 0.3 mol% of lithium concentration with (low standard error) value. The TL glow curve for Al_2O_3: Li (0.1 to 0.5% mol concentration) was also studied. From the results, different concentration of Li playing a main role in production of TL glow curve. The prominent glow peak observed at high temperature around 175°C. This high glow curve temperature due to high-energy traps and it is very useful for TLD phosphor characteristics [28]. No such work for Al_2O_3: Li has been reported so far, but historically, the first basic TL material was LiF:Ti, Mg (TLD 100), whose properties are still being studied extensively. Interest of this material is due to its human tissue equivalence (Z_{eff} = 8.04) which is an important factor for personal dosimetry, simple glow curve, high sensitivity and limited fading over a certain research.

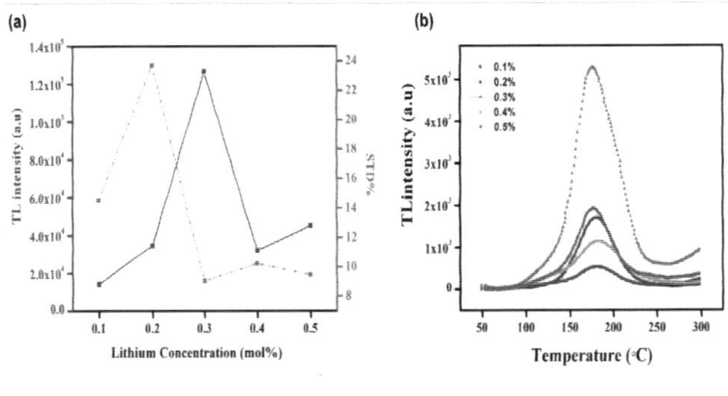

Fig 3. (a) TL intensity and correspond standard deviation as a function of different Li concentration mole (0.1 to 0.5%) (b) Glow curve recorded at different concentration mole (0.1 to 0.5%) of Li dopant at a constant exposed to dose 50Gy

3.4 Determination for the optimum germanium dopant concentration

Figure 4(a) represents the TL intensity and corresponding to standard deviation of Al_2O_3 at different Ge concentration. The TL intensity of the sample was found to vary with increasing Ge concentration. The TL intensity increase from 1 mol% Ge to maximum intensity at 3mol% Ge concentration 349747.04 (a.u). Then the intensity decreases with further increase of Ge concentration. This may be due beyond 3 mol% concentration, quenching is bound to occur hence the optimum concentration for Ge only suite for this concentration. Highest TL intensity at 3 mol% Ge is attributed to the recombination of the excited electrons from the valences band with defects produced via irradiation [29]. As a conclusion, it is found that the optimum concentration value of Ge dopants is at 3 mol% with very low standard error. At the same Figure 4(b) shows the glow curves Al_2O_3: Ge samples for different concentration mole. Two peaks are observed, one prominent peak at around 175°C and another hump around 230°C. It is apparent that the TL intensity of the TL peak arrive the maximum value at 0.3% mol. For further study, it is found that the optimum concentration value of Ge dopants is at 0.3 mol% with very low standard error.

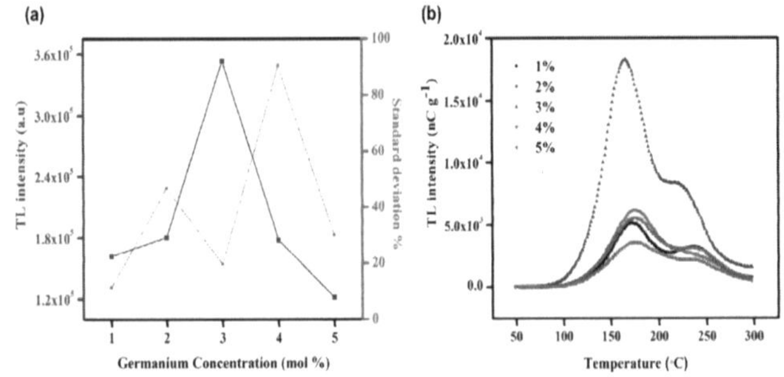

Fig 4. (a) TL intensity and correspond standard deviation as a function of different Ge concentration mole (1 to 5%) (b) Glow curve recorded at different concentration mole (1 to 5%) of Ge dopant at a constant exposed to dose 50Gy

3.5 Comparison of TL glow curve for each sample

Figure 5 represents the glow curve of materially from each optimize concentration as mention in section 3.2 to 3.5. A very strong peak appears at 175°C for all samples but a small hump also appears for Ge sample at approximately 230°C. Based from the comparison of these three dopant materials, it can be observed that Al_2O_3 doped 0.3% Germanium gives the highest TL intensity compared to other dopants which is 93 times higher. The highest TL intensity that is reflected from either the total integrated area under glow peaks or that recorded in the reading system is detected at 3.0mol% Ge. Simple glow curve is another important property of a dosimeter. A good TLD material should display a simple glow curve and if possible has a single peak at around 200°C [30]. Furthermore, it should also have high resistance to environmental factor, as well as negligible dependence of its emission on radiation energy. As a conclusion for further study Al_2O_3: Ge at 3.0mol% was selected to study the glow curve of germanium dopant properties.

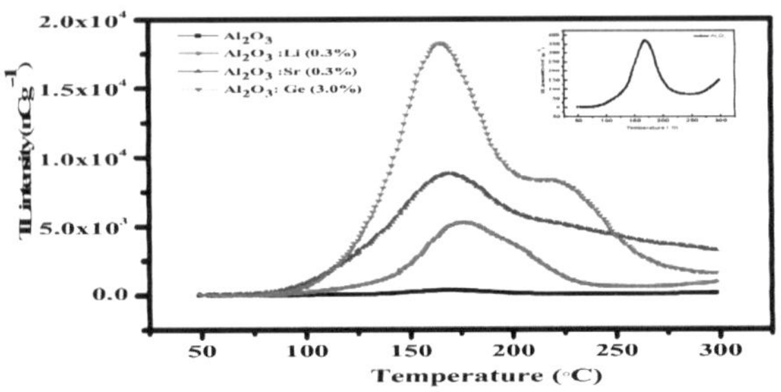

Fig 5. Glow curve recorded at different dopant material from optimum each concentration at a constant exposed to dose 50Gy

DOI: 10.1051/epjconf/201715600001

3.6 Glow curve analysis

Figure 6 shows the TL glow curve of aluminium oxide doped with 3.0 mol% germanium after being irradiated to Cobalt-60 gamma rays at dose ranging from 10Gy to 80 Gy. The glow curves recorded for all doses delivered the glow peak falls at around 175 °C and 230°C. The figure also indicates the TL intensity increase as the delivered dose increase. This results also agrees when α- Al_2O_3 doped either with Tb^{3+}or Tm^{3+} was prepared by combustion synthesis techniques studied by [11]. In this method the reactants (aluminium nitrate, urea and terbium or thulium nitrate) are ignited in a muffle furnace at temperatures as low as 500°C. The samples were annealed at temperatures ranging from 1000 to 1400°C for 4h in order to obtain the pure α-phase structure and were then irradiated with a Co-gamma radiation source. Tb^{3+} gives clearly TL peaks at around 190°C after sintering the sample at 1400°C. The dose response for 0.1 mol% for Tm^{3+} gives TL peaks around (~190°C) but the TL intensity increased 4 times better using Tm^{3+}. Another work also by [31], tried to improve tld material nanoparticles for Al_2O_3 by different element like Eu, Tb, Dy, Cu and Ag were synthesized by chemical combustion (propellent) method. The samples were exposed to 100Gy of gamma-rays from a ^{137}PCs source. All the doped samples show poor TL sensitivity with two humps at around 150°C and 260°C except for Tb dopant. The glow curve for Tb has three major peaks; the highest is located at around 370°C, second 230°C and third almost two components located at around 125°C and 90°C.

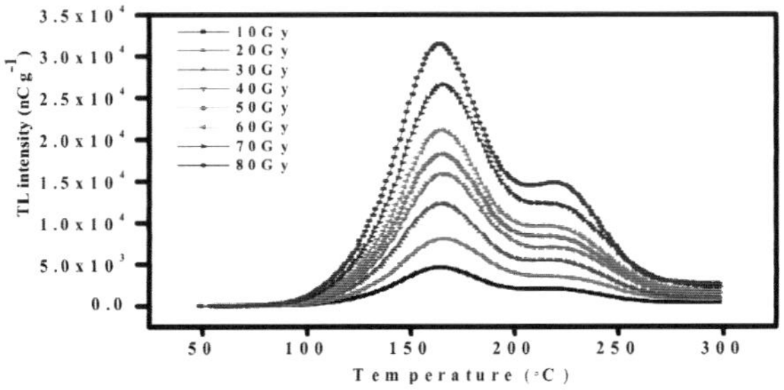

Fig.6. TL glow curve of 3.0 mol% germanium doped aluminium oxide at different doses of Cobalt-60 gamma rays

3.3.2 Linearity

Linearity of dose response is one of the most crucial properties of a TL phosphor. Figure 7 indicates TL intensity versus temperature for Al_2O_3: Ge 3% mol exposed to Cobalt-60 Gamma Rays at various doses ranging from 10-80 Gy. The linear response indicates that the TLD is directly proportional to the dose of radiation it was irradiated with. It clearly reveals a good linear dose response with the delivered dose up to 80Gy. Regression coefficient (R^2) for Al_2O_3: Ge 3% is 0.9903. From the figure, the sensitivity of Al_2O_3: Ge 3% is 10019 nCg^{-1} Gy^{-1}.

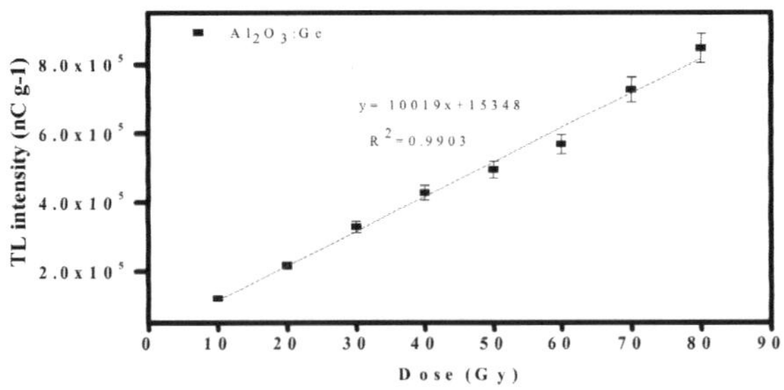

Fig7. TL intensity of Al_2O_3: Ge 3% mol irradiated to Cobalt-60 gamma ray for doses ranging from 10Gy to 80 Gy

Conclusion

From the results,it is possible to conclude that the CS method is very suitable for the preparation of Al_2O_3 doped materials for dosimetric applications. This technique is quick and easy process, with as main advantages the saving time and energy. From the comparison of three doped materials, Ge, Sr and Li, the concentration analysis show that the Ge concentration that optimizes the TL response 3mol% and in order to obtain well TL response peak for dosimetric applications. The crystallinity of the samples is proven by the XRD sharp peaks. The results will provide TL intensity, which depends on the delivered dose. The sensitivity found for Al_2O_3 : Ge 3% is 10019 $nCg^{-1}\ Gy^{-1}$

Acknowledgement

The authors would like to thank the Ministry of Education (MOE) of Malaysia for the financial support (Vote Number: 10H20) and Universiti Kebangsaan Malaysia for providing irradiation source of Cobalt-60 gamma Cell Model 220 Excel and Universiti Teknologi Malaysia for providing research facilities.

References

[1] Konrad, A., Fries, T., Gahn, A., Kummer, F., Herr, U., Tidecks, R., Sammer, K., 1999. Chemical vapor synthesis and luminescence properties of nanocrystalline cubic Y_2O_3. Eur. J. Appl. Physiol. 86, 3129-3133.

[2] O. Margit., T. BIRO., Aluminium oxide in TL Dosimetry. Nucl. Instr. Meth, 175, 60-61, (1980).

[3] M.S., Akselrod, A.E., Akselrod, S.S., Orlov, S. Sanyal and T.H., Unerwood, FluorescentAluminium Oxide Crystals for Volumetric Optical Data Storage and Imaging Applications. Journal of Fluorescence.13,6. (2003).

[4] A.J. J, Bos, High sensitivity thermoluminescence dosimetry. Nucl. Inst Meth. Phys. Res. B. 184: 3-28, (2001).

[5} D.V, Barros, S.M. de Azevedo Walter, M. Khoury and Pedro Linhares Filho, Combustion synthesis: A suitable method to prepare Al_2O_3 doped materials for thermoluminescent dosimetry. Radiat. Meas. 43, 345-348, (2008).

[6] M.E.A. Andrade, W.M. Azevedo, V.S.M Barros, and H.J Khoury, Thermoluminescence of aluminium oxide co-doped with terbium and thulium obtained via combustion synthesis. Radiat. Meas. 46, 1474-1476, (2011)

[7] J.V. Soares, C.F, Gugliotti, Y.S. Kawashima, S.H. Tatumi, J.C.R. Mittani, Thermoluminescence and optically stimulated luminescence characteristics of Al_2O_3 doped Tb. Radiat.Meas, 71, 78-80, (2014).

[8] A.A. Kaplyankii, A.B. Kulinkin, A.B. Kutsenko, S.P. Feofilov, R.I. Zakharchenya, T.N. Vasilevskaya, Optical spectra of triply-charged rare-earth ions in polycrystalline corundum. Phys. Sol. State, 40, 1310-1316, (1998).

[9] N. Can, P.D. Townsend, and D.E. Hole, Enhancement of luminescence by pulse laser annealing of ion-implanted europium in sapphire and silica. J. Appl. Phys.78, 6737-6744, (1995).

[10] J. Azorin, Preparation methods of thermoluminescent materials for dosimetric applications: An overview. Appl. Radiat. Isot. 83, 187-191, (2002).

[11] G. Hirata, N. Peres, M. Tajeda, J.A. Gonzalez-Ortega, and J. McKittrick, Luminescence study in Eu-doped aluminium oxide phosphors. Opt. Mater. 27,1311-1315, (2005).

[12]V.SM. Barros, W.M.de Azevedo, H.J. Khoury, M.E. A. Andrade, and P. Linhares Filho, Thermoluminescence study of aluminium oxide doped with terbium and thulium. Radiat. Meas. 45,435-437. (2010).

[13] B.N. Lakshminarasappa, J. R.Jayaramaiah, & B. M. Nagabhushana, Thermoluminescence of combustion synthesized yttrium oxide. Powder Technol, 217: 7–10. (2012).

[14]A.Sharma, A.Rani, A.Singh, O. P. Modi, & G. K. Gupta, Synthesis of alumina powder by the urea–glycine–nitrate combustion process: a mixed fuel approach to nanoscale metal oxides. Applied Nanoscience, 4(3): 315–323, (2014).

[15] R. K. Lenka, T. Mahata, P. K. Sinha, & A. K. Tyagi, Combustion synthesis of gadolinia-doped ceria using glycine and urea fuels, J Alloys Compd, 466: 326–329. (2008).

[16] Ravichandran, A.T., Catherine, K., Pushpa, S., Ravichandran, K., Karthika, K., Nagabhushana, B.M., Mantha, S., and Swaminathan, K. Effect of Al doping on the structural and optical properties of ZrO_2 nanopowders synthesized using solution combustion method. Superlattices and Microstructures.75, 533-542. (2014).

[17] 16. R. K. Lenka, T. Mahata, P. K. Sinha, & A. K. Tyagi, Combustion synthesis of gadolinia-doped ceria using glycine and urea fuels, J Alloys Compd, 466: 326–329. (2008).

[18]M. C. Gardey Merino, G. E. Lascalea, L. M. Sánchez, P. G. Vázquez, E. D. Cabanillas, & D. G. Lamas. Nanostructured aluminium oxide powders obtained by aspartic acid-nitrate gel-combustion routes. J Alloys Compd, 495(2):578–582. (2010).

[19]F.O. Ogundare, S.A. Ogundele, M.L. Chithambo and M.K. Fasasi, Thermoluminescence characteristics of the main glow peak in α- Al_2O_3:C exposed to low environmental-like radiation doses. J.Lumin. 139, 143-148. (2013).

[20] Kar, S., Debnath, C., Verma, S., Dhamgaye, V. P., Lodha, G. S., & Bartwal, K. S. Thermoluminescence studies on single crystal, polycrystalline and glass lithium tetraborate samples irradiated by X-rays from Indus-2. Physica B: Condensed Matter, 456, 1–4, (2015)

[21] V.T. Adamiv, Ya.V. Burak, I.V. Kityk, J. Kasperczyk, R. Smok, M. Czerwihki, Nonlinear optical properties of $Li_2B_4O_7$ single crystals doped with potassium and silver, Opt. Mater. 8 207–213, (1997)

[22] S. Kar, K.S. Bartwal, Growth optimization of $Li_2B_4O_7$ crystals and their characterization, Cryst. Growth Des. 7 2522–2525, (2007)

[23] D.S., Khatri, Ayush Khare, Piyush Jha., Thermoluminescence studies of $SrAl_2O_4$: Eu Phosphors at different by concentrations, Chalcogenide Letters, 10, 121-129, (2013)

[24] E. Pekpak, A.Yilmaz, and G. Ozbayoglu. An Overview on Preparation and thermoluminescence Characterization of Lithium Borates for Dosimetric Use. The Open Mineral Processing Journal. 3, 14-24. (2010).

[25] F.O. Ogundare, S.A. Ogundele, M.L. Chithambo and M.K. Fasasi, Thermoluminescence characteristics of the main glow peak in α- Al_2O_3:C exposed to low environmental-like radiation doses. J.Lumin. 139, 143-148. (2013).

[26] M. Farahmandjou, N. Golabiyan, Solution Combustion Preparation of Nano-Al_2O_3: Synthesis and Characterization, Trans. Phenom. Nano Micro Scales. 3(2): 100-105. (2015)

[27] Abha H. Oza, N.S. Dhoble, K.Park and S.J.Dhoble, Synthesis and thermoluminescence characterizations of $Sr_2B_5O_9Cl:Dy^{3+}$ phosphor for TL dosimetry, The Journal of Biological and Chemical Luminescence,30, 768-774. (2015)

[28] S. P Puppalwar, S. J Dhoble, N. S Dhoble, & A. Kumar, Nuclear Instruments and Methods in Physics Research B Luminescence characteristics of Li 2 NaBF 6: Cu phosphor, *274*, 167–171, (2012).

[29] E. Pekpak, A.Yilmaz, and G. Ozbayoglu. An Overview on Preparation and thermoluminescence Characterization of Lithium Borates for Dosimetric Use. The Open Mineral Processing Journal. 3, 14-24. (2010).

[31] Numan Salah., Najlaa Alharbi, D., Sami Habib, S., and Lochab, S.P. Thermoluminescence properties of Al_2O_3: Tb nanoparticles irradiated by gamma rays and 85 MeV C^{6+} ion beam. Journal Luminescence. 167,59-64. (2015).

Thermoluminescent properties of Dy doped calcium borate based glass for dose measurement subjected to photon irradiation

H. A. Tajuddin[1] , *W. M. S.* WanHassan[1], *S. F.*, Abdul Sani[2], *Nurul Syazlin* Shaharin[1]

[1] Department of Physics, Faculty of Science, Universiti Teknologi Malaysia, 81310 UTM Johor Bahru, Johor Darul Takzim, Malaysia.
[2] Department of Physics, University of Malaya, 50603 Kuala Lumpur, Malaysia.

Abstract. This study presents the thermoluminescent (TL) dosimetric properties of calcium borate glass with various dopant concentration of dysprosium (Dy). Calcium borate glass is a new potential material to be used in radiation measurement with absorption coefficient that is close to human bone. A series of glasses based on chemical equation xCaO-(100-x) B_2O_3 system, x = 0.1, 0.2, 0.3, 0.4, 0.5 ($0\leq x \leq 100$) % weight have been prepared by melt quenching method. The X-ray diffraction analysis of glass samples were carried out and the result showed a broad peak, which confirmed the amorphous nature of the glass. The $70B_2O_3$-30CaO glass sample was found as the most stable among other glass samples studied. Present work focuses on $70B_2O_3$-30CaO glass of (0.01-0.4) mol% Dy-doped in order to investigate the thermoluminescence (TL) properties, in particular, dose-response and fading. The glass samples were irradiated to dose range of 0.5-4.0 Gy subjected to 6MV photon irradiations of LINAC Primus MLC 3339. TL response of 0.3 mol% Dy-doped $70B_2O_3$-30CaO glass was found to produce highest response, with good linear dose-response relationship.

Keywords: Thermoluminescence, Borate glass, Dopant, Dysprosium, Photon irradiation.

1 Introduction

Since 1950s there have been extensive research on thermoluminescence dosimeter (TLD) and its application in radiation dosimetry. These include LiF and CaF_2, the sulphates groups $MgSO_4$, $CaSO_4$ and $BaSO_4$, and the oxides group AL_2O_3, BeO and SiO_2. A current highly popular medium for radiation dosimetry are the phosphor-based thermoluminescent dosimetry (TLD) family, of which Li:Mg,F,Ti is perhaps the most well known. Problems arise when such TLDs are used for radiation dose evaluations in radiotherapy and environmental monitoring, not least because of their relatively large dimensions (typically

*Corresponding author: hamizaahmadtajuddin@gmail.com

some several mm and more) and hygroscopic nature, the latter certainly limiting their versatility. Borate glass compounds have found to be advantageous materials for solving these problems and widely investigated for radiation dosimetry application [1]. It was recognised that pure borate also known as alkali borate glasses have attractive interest as thermoluminescence (TL) dosimeter material, providing the effective atomic numbers approaching to that of human tissue (Z_{eff}=7.42) with very low hygroscopic nature [2-3]. Given that the glass TL dosimeters are resistant to water and therefore overcoming the adverse effect on the TL efficiency of the material that associated with non-radioactive relaxations during thermal stimulation [4].

In present study, it is intended to investigate new compound of calcium borate (CaB_4O_7) glasses for application as thermoluminescence dosimeters (TLD). It is noted that calcium borate glasses have good chemical stability that of an important consideration for developing alternative forms of dosimeter [1, 5-8]. As reported elsewhere, the luminescence spectra and the glow peak in TLDs arise mainly from the signal in the doped core [21]. Rare earth elements as dopant in silica materials have gained much interest of many researchers as a result of 4f electronic configuration of rare earth in various glass matrixes and therefore leads to emission of ultraviolet and infrared [9], a matter which can be used to great benefit of radiation dosimetry system. The work presented herein makes use of the TL signal stored in rare earth based Dy_2O_3 as dopant in the calcium borate glass subjected to ionising radiation, offering an excellent TL yield and dose sensitivity [10]. The latter is considerably enhanced by the presence of structural defects in the material that have been created by the doping process.

2 Materials and Methods

A total of six glass samples were prepared from high grade purity chemical substances in the form of powder using conventional melt quenching method. The chemical compounds used in this work were B_2O_3 (69.6217 $gmol^{-1}$), $CaCO_3$ (56.08 $gmol^{-1}$) and Dy_2O_3 (372.99 $gmol^{-1}$) with 99.5%, 99.9% and 99.9% purity, respectively. In order to obtain 30CaO-(70-x) B_2O_3 doped with Dy ($0 \leq x \leq 0.4$) mol % glass sample, the chemicals were weighted using an analytical balance and subseqeuntly mixed in alumina crucible. The glasses were melted in an electric furnace at 1200˚C for 30 minutes (later being called as melt). The glasses were quenched by pouring out the melt on a flat metal plate and pressed by different metal plate with thickness of 1.5-2 mm. The samples were then annealed for 2 hours at 300˚C before being cooled at room temperature. The above steps were repeated for preparing other samples. Table 1 shows the details of the samples prepared.

Table 1 Details of the samples.

Sample No	**Remark**[a]	**Nominal Composition (mol %)**		
S1	Colorless glass	H_3BO_3	$CaCO_3$	Dy_2O_3
S2	Colorless glass	70	30	0.0
S3	Colorless glass	69.95	30	0.05
S4	Colorless glass	69.9	30	0.1
S5	Colorless glass	69.7	30	0.3
S6	Colorless glass	69.6	30	0.4

[a] Visual inspection

The vitreous phase of the samples was characterised through X-ray diffraction (XRD) (Siemens D5000, Canada) technique. It is equipped with diffraction software analysis of Cu K-α radiation, operating at 40 kV, 30 mA and Bragg-Brantano geometry at room temperature. Energy dispersive X-ray(EDXS) analysis (model JSM-6701, UK Company) was performed to measure the elemental composition of the glass samples investigated herein. Effective atomic number (Z_{eff}) results were determined by using the experimental values of weight fraction and compared to the theoretical values.

The glass samples were exposed to 6 MV photon, use being made of linear accelerator (LINAC) Primus Multileaf Collimator 3339 MLC located at the Department of Radiotherapy and Oncology, Hospital Sultan Ismail in Johor Bahru. The photon irradiation was performed in the range of doses of 0.5 Gy to 4.0 Gy with dose rate of 200 MU/min. During irradiation, a bolus material was placed on top of the perspex to modulate the range of dose from external beam of radiation. The source to sample distance (SSD) was set at 100 cm, with a field size of 15 x 15 cm^2. In all cases, the fibres were placed at the centre of the field. The glass samples were read 24 hours after irradiation, this would reduce the effect of spurious thermoluminescence signals. Once the samples rested for 24 hours, the TL glow curve measurement were carried out using a TLD reader (model 4500, Harshaw Chemical Company USA) provided by Nuclear Laboratory, Department Physics, Faculty of Science, Universiti Teknologi Malaysia (UTM). During readout, the heating rate used was constant at 10 °C s^{-1} and a continuous flux of N_2 is supplied to reduce chemiluminescence in the heated samples.

DOI: 10.1051/epjconf/201715600002

3 Results and discussion

3.1 X-ray diffraction (XRD) analysis

The XRD spectra of the samples of various Dy dopant concentrations is illustrated in Fig. 1. According to XRD data, it was confirm that the samples revealed bands without any discrete or continuous sharp peaks which indicated the amorphous character of the samples. This patterns were coincident with the card reported by the joint committee on JCPDS no: 76-0640.

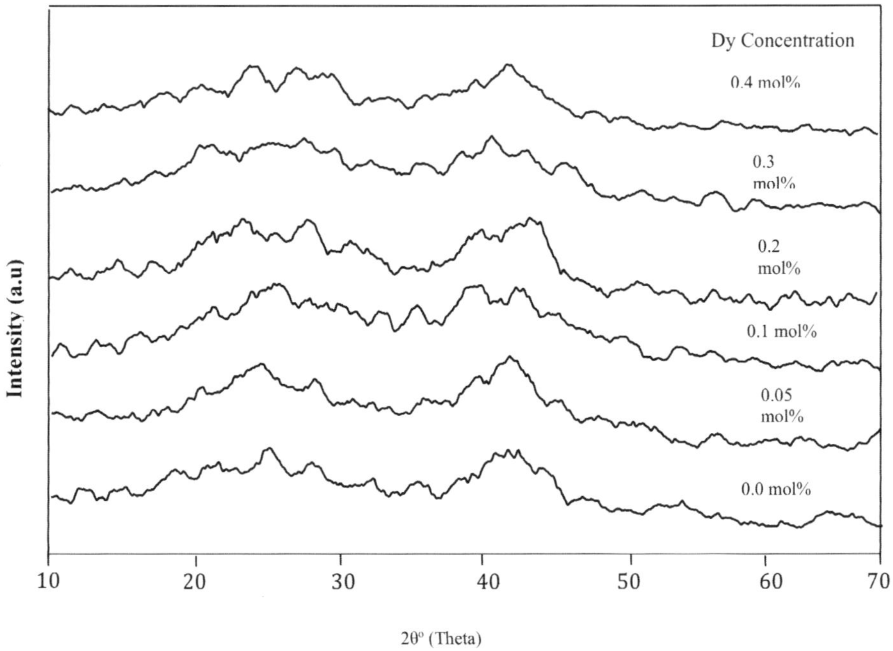

Fig. 1. The X-Ray diffraction spectra for Dy- doped and undoped glass samples. The dopant concentrations of Dy are labelled on the right hand side of the figure.

3.2 Thermoluminescence (TL) Properties

3.2.1 Optimum dopant concentration of Dy_2O_3

In this research, 30CaO-(70-x) B_2O_3 glass system has been doped with 0.1, 0.2, 0.3, and 0.4 mol% dysprosium (Dy) and irradiated at 6 MV photon with an entrance dose of 4 Gy. A graphical representation of the total number of counts recorded against temperature is

DOI: 10.1051/epjconf/201715600002

as seen in Fig 2. A typical pattern for the glow curve obtained from the glass samples are a single broad peak, being characteristic of amorphouse media and does not change appreciably depending upon the concentrations. After the irradiation process, it was found that 0.30 mol% Dy-doped calcium borate glass sample resulted highest TL response, being selected for subsequent measurements in investigating the response of the sample for different dose ranges, sensitivity, fading and reproducibility studies. It is also apparent that the glow peak of undoped calcium borate glass exhibits at 267°C, being significantly elevated as the dopant concentration increased at 286°C, in particular, 0.3 mol % of Dy-doped glass sample. It might be expected that the TL intensity will increase with dopant concentration, the number of electron traps and recombination centres increasing with dopant concentration, however, beyond an optimised concentration (nominal 0.30 mol%), the TL intensity of the samples actually decreases. This occurs because the distance between traps reduces as the number of traps and recombination centres increases, raising the probability for the light emitted from one recombination process to be absorbed by an electron in another trap. This phenomenon is called self-absorption. Moreover, the well known concentration quenching (CQ) effect have also attributed to this behaviour, which theoretically treated by Johnson and Williams (1950) [11] and later on modified by Ewles and Lee (1953) [12].

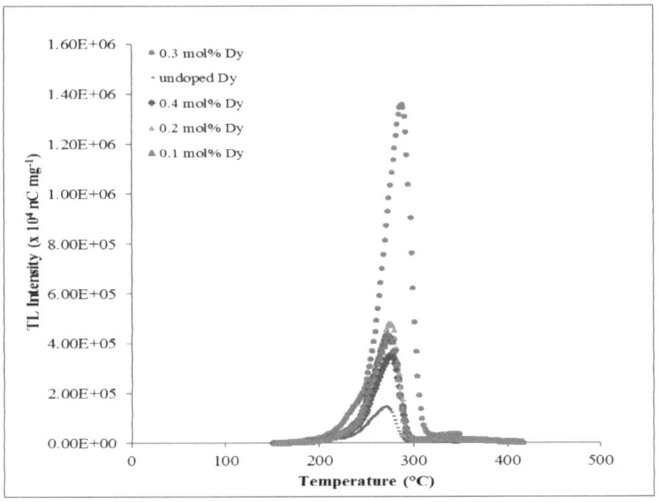

Fig. 2. Glow curves for different mol (%) dopant concentration.

 DOI: 10.1051/epjconf/201715600002

3.2.2 Optimum Annealing Temperature

To investigate suitable annealing temperature for the selected calcium borate glass (nominal 0.30 mol% of Dy-doped), four different annealing temperatures were tested, performing at 100, 200, 300 and 400 °C for a period of 1 hour and in all cases irradiated by megavoltage photon beams of 6 MV at a dose of 4 Gy. Annealing process is crucial before making any irradiations and subsequent TL measurements in prior to achieve highest reproducibility for both background and TL signal. According to Prokic [13], this process also allows to eliminate the effect of previous irradiation. The TL intensity of the 0.3 mol% Dy-doped glass sample against annealing temperature was measured and the summary results are shown in Fig. 3. In present study, each data point was obtained by taking an average of three individual measurements in order to plot the standard deviation against the annealing temperature. TL intensity with lowest standard deviation of the annealing temperature was found to be at 300°C, thus selected for further TL characteristic studies.

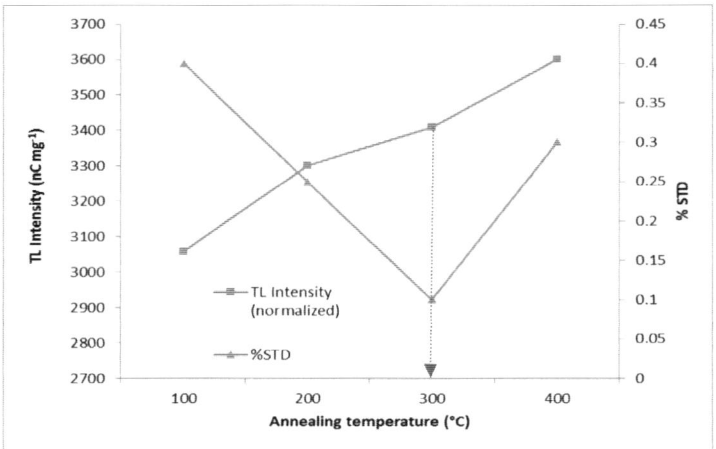

Fig. 3. The TL response and the corresponding standard deviation as a function of the annealing temperature.

3.2.3 Optimum Heating Rate

It is important in use of the calcium borate glass sample that heating rate be characterised. The heating rate is a dynamic parameter which influences the characteristics of TL glow curve. As the heating rate increases the peak of glow curve will shifts to higher temperatures and the TL intensity measured will be decreased [14-15]. After exposure to 6 MV photon of 4 Gy dose, the glass samples were readout at five different heating rates of 2, 5, 10, 12 and 15 °C s^{-1}. The optimum heating rate is obtained at 10 °C s^{-1}, considering the highest TL intensity with lowest standard deviation (see Fig. 4).

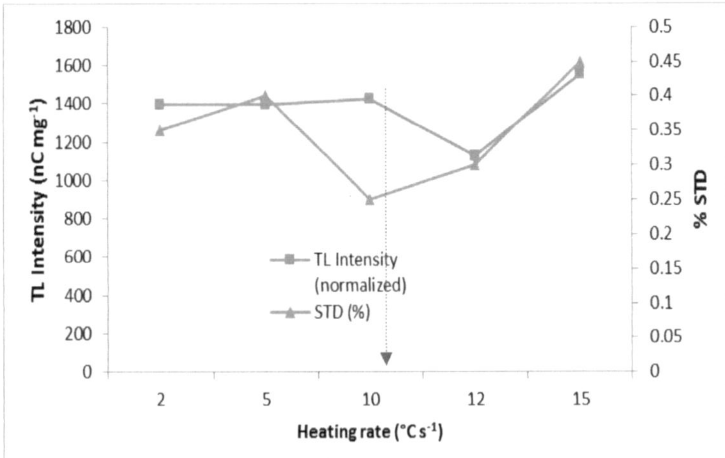

Fig. 4. TL response (normalized to unit mass) and the corresponding standard deviation as a function of the heating rate.

3.1.4 Glow Curve

Glow curve is obtained by measuring the luminescence and plot the results as a function of temperature. The relative amplitudes of the peaks indicate approximately the relative population of trapped electrons and trap depth, provided the heating rate is constant through the glow curve. Fig. 5 shows TL growth patterns of the prominent peaks of Dy-doped calcium borate glass samples with increase of absorbed doses of 2, 3 and 4 Gy. It was observed that the glow peaks were gradually enhanced with increased in absorbed dose delivered to the glass sample. This can be explained by the number of filled traps, corresponding to the area under the glow peak, which in turn represented the amount of radiation energy deposited into the glass sample. These traps have consequently enhanced the TL intensity as a result of extrinsic doping of Dy, causing the glow peak shifted to higher temperature from 301.65°C to 309.94°C with respect to increment in absorbed dose. This behaviour is caused by releasing the deeper traps of the charge carriers in glass host so in principle, the concentration and storage capacity of these traps can provide a medium suitable for dosimetric application [16].

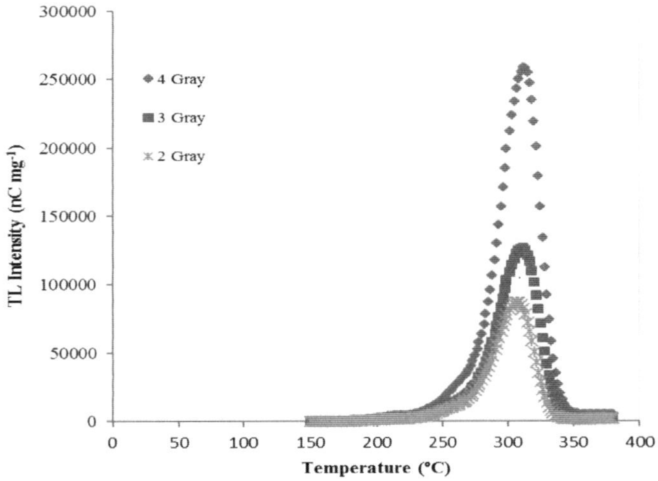

Fig. 5. Glow curve of 0.3 mol% dysprosium doped calcium borate glass at different doses of photon irradiation.

3.1.5 Linear Dose Response

One of the important characteristics to assess the performance of dosimeters is through the study of linearity. It is desirable for TL materials to possess a linear relationship between the TL emission to the absorbed dose [17-20]. McKinlay (1981) has reported that the TL response of well-known phosphor TLDs, such as TLD-100 is usually, but not exclusively linear at low absorbed dose values, becoming supralinear and finally saturating at higher values. In this work the linearity for TLD-100, Dy-doped and undoped calcium borate glass samples subjected to 6 MV and 10 MV photons irradiations for a wide dose range (0.5 Gy - 4 Gy) were examined. Figure 6 shows the TL-dose relationship for each sample. Each data point represents the average of three individual glass sample readings together with the standard error of the mean. The increase in TL yield with dose remains linear over a wide range of values for all types of samples.. However, both glass samples produce lower TL response compared to that of TLD-100. A regression coefficient (R^2) is another indication for linearity between dose and TL response. The solid lines are least square fits to the data, obtaining correlation coefficients of 0.9916, 0.9831 and 0.5285, respectively for TLD-100, Dy-doped and undoped glass samples. It is evident that the Dy-doped glass samples provide the basis for sensitive dosimetry throughout this range to that undoped glass sample, the TL intensity for the irradiation conditions increasing by 5% per additional 1 Gy of dose delivered.

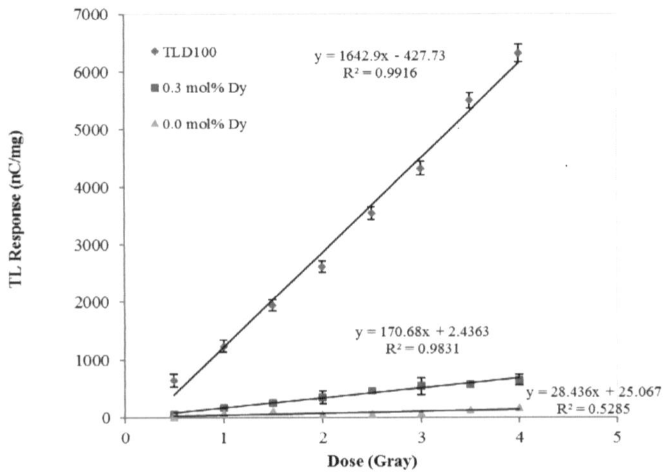

Fig. 6. The dose response for 0.3 mol% Dy-doped and undoped calcium borate glass samples, in comparison with TLD-100 irradiated to a dose of 0.5 Gy to 4 Gy with MV photon beams of 6MV.

3.1.7 Fading

Fading is the loss of stored thermoluminescene signals and depends largely upon the depth of traps and the storage temperatures of the TL materials. In general, if the TL signal from TLD sample is found to be unstable over the time after exposure to radiation, the signal is said to have faded. Note that fading can be affected by the heat treatment applied during annealing, readout procedures and exposure to visibile radiation and ultraviolet radiation. In present study, the fading effect of 0.3 mol% Dy-doped glass sample was observed over a period of 60 days in order to understand the loss of TL response post irradiation. The percentage losses of TL signals can be obtained from equation 3.1 below (Suhairul, 2009):

 DOI: 10.1051/epjconf/201715600002

$$Percentage\ loss\ of\ TL\ sign = \frac{TL\ response\ for\ 1st\ day - TL\ response\ for\ last\ day}{TL\ response\ for\ 1st\ day} \quad (3.1)$$

The fading effect of Dy-doped calcium borate glass, undoped calcium borate glass and TLD-100 over a period of 60 days storage time for photon beam irradiations of 6 MV are demonstrated in Fig. 8, and 10, respectively. After 60 days, the average results are tabulated in Table 2. TLD-100 shows minimal loss in TL intensity of 8.7 % followed by Dy-doped of 13.97 % and undoped sample was observed to have maximum fading of 53.96 % after 60 days. Table 3 shows different type of TLD with their own thermal fading.

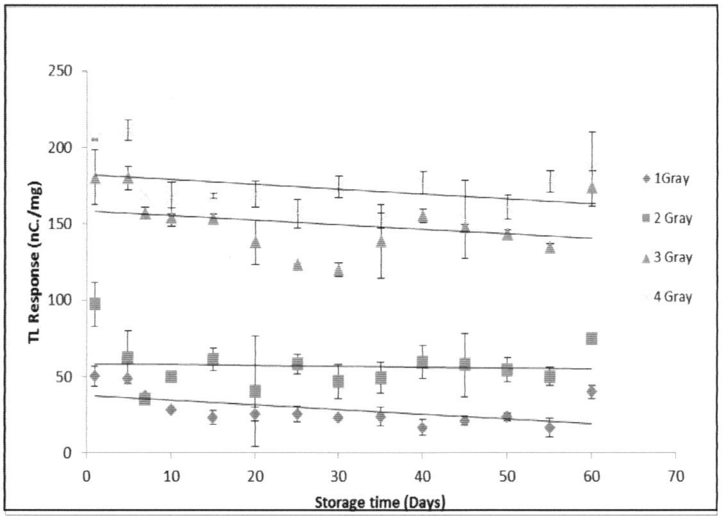

Fig.8 60 days variations of the TL response of the 0.3 mol% dy doped calcium borate glass

 DOI: 10.1051/epjconf/201715600002

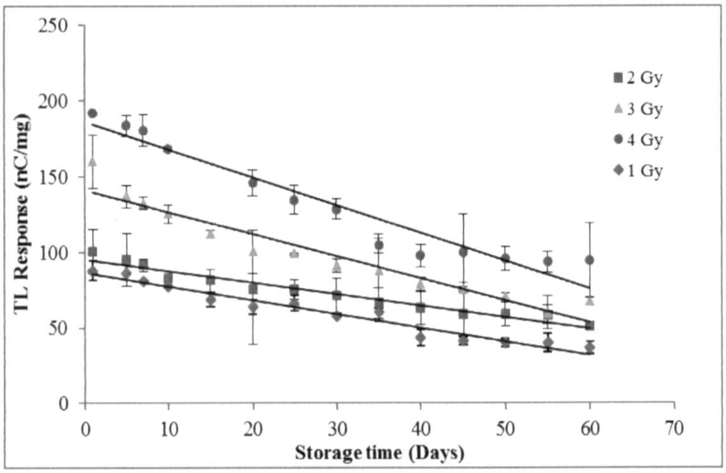

Fig. 9 60 days variations of the TL response of the undoped calcium borate glass

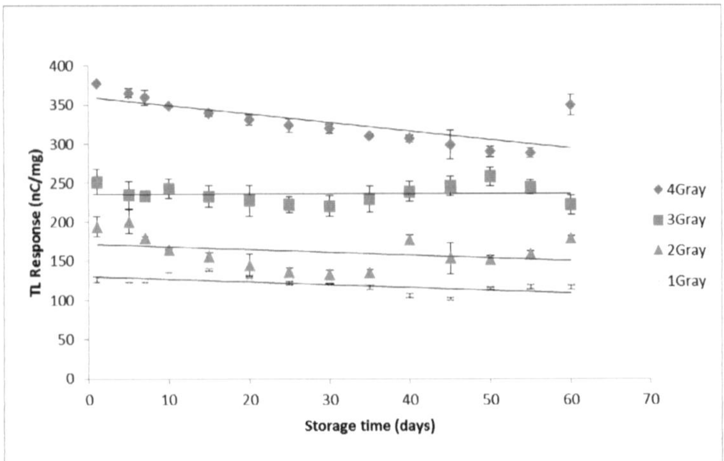

Fig. 10 The fading of TLD-100 for 10 MV photon irradiation

Table 2 Summary of percentage loss of TL signal (due to fading) over the first 24 hours post-irradiation for various TL materials.

TL Materials	Dose (Gray)				Average
	1	2	3	4	
Calcium Borate: Dy	19.97	23.19	3.45	9.28	13.97
Undoped Calcium Borate	58.15	49.08	58.04	50.58	53.96
LiF: Mg.Ti (TLD-100)	8.13	7.66	11.64	7.39	8.70

DOI: 10.1051/epjconf/201715600002

Table 3: Fade rates for some commercial TLD materials

TLD Dosimeter	TLD Materials	Thermal Fading
TLD 100	LiF:Mg, Ti	5 – 10% per year
TLD 200	CaF_2:Dy	16% in two weeks
TLD 400	CaF_2:Mn	15% in three months
TLD 500	Al_2O_3:C	3% per year
TLD 600	^{6}LiF:Mg, Ti	5% per year
TLD 700	LiF:Mg, Ti	5% per year
TLD 800	Li_2B4O_7:Mn	5% in three months
TLD 900	$CaSO_4$:Dy	8% in six months

3.1.8 Sensitivity

According to Furetta, 2003 [23], sensitivity of TLD can be defined as TL response (in reader unit) per unit of dose and unit of mass of the sample. It also can be expressed following equation;

$$S = \frac{TL}{D.m} \tag{3.2}$$

where TL is the thermoluminescence response measured by the TLD reader, *D* is deposited dose and *m*, mass of the sample. Sensitivity also depends upon the readout system used in the measurement, for instance, heating rate and light detection system. Table 3 shows the average ratio of sensitivity for TLD-100, Dy-doped and undoped calcium borate glass sample subjected to 10 MV and 6 MV photon irradiations. Comparison has been made with TLD-100, producing greater sensitivity and outperform other materials. Taking TLD-100 as the benchmark, Dy-doped calcium borate glass exhibits lower sensitivity of some 5 times to that TLD-100. However, as expected, the doping presence has produce a significant sensitivity to that undoped glass sample of some 3 times. The significant energy dependency of the doped calcium borate glass in the MV range point to a potential for energy discrimination of the incident photons if these glass samples are used in conjunction with TLD-100.

Table 4 TL sensitivity (TL yield.mg^{-1}.Gy $^{-1}$) for TLD-100, Dy- doped and undoped Calcium Borate glass sample for 6 MV and 10 MV photon irradiations

Materials	Energy/Dose	1 Gy	2 Gy	3 Gy	4 Gy	Average
LiF: Mg.Ti (TLD-100)	10 MV	79.8	93.5	102.3	206.3	120.475
	6 MV	51.2	54.2	59.6	65.2	57.5
Calcium Borate: Dy	10 MV	21	26.5	31.5	43.5	30.625
	6 MV	35.2	28.1	42.0	35.9	35.3
Calcium Borate	10 MV	8.1	5.3	5.1	4.9	5.85
	6 MV	2.3	1.6	2.0	2.0	2.0

3.1.8 Effective atomic number, Z_{eff}

The response of a TL material to x- and gamma rays detection depends on the atomic number of its constituents as in [18]. The effective atomic number, Z_{eff} of an ideal dosimeter is very important in radiotheraphy for calculating the treatment dose in cancer patients [16], defining an equivalent value for mixtures and compounds that offers Z_{eff} equal to or close to the composition of human biological tissue. The Z_{eff} of glass sample used herein was measured through the strength of different interaction probabilities or cross section between gamma rays and the studied materials [24] prior obtaining its expected TL response at different energies, use being made of XCOM software method. 30CaO-(70-x) B_2O_3 glass system (where x = 0, 0.05, 0.1, 0.2, 0.3, 0.4 and 0.5 mol % of dopant concentration) was initially converted to weight fraction and the Z_{eff} value a found to be in the range of 12.14 to 13.22, specifically for x = 2.94 mol% and x = 3.5 mol%, respectively. Also calculated the percentage difference to be 8% with average of 12.68. The variation of the mean values of Z_{eff} for the present Dy doped glass system is shown in Fig 11. It is apparent that Z_{eff} increases with increasing Dy concentration and the values ranging from 12.68 to 15.13. This variation is represented by an excellent correlation coefficient, R^2 = 0.98. These results indicate that Dy-doped calcium are not tissue equivalent (value of Z_{eff} in soft tissue is 7.5) and as such the assessment of dose deposition in tissue would need to be corrected and calibrated. However, the effective atomic number is not greatly different to that of the bone, 13.8.

DOI: 10.1051/epjconf/201715600002

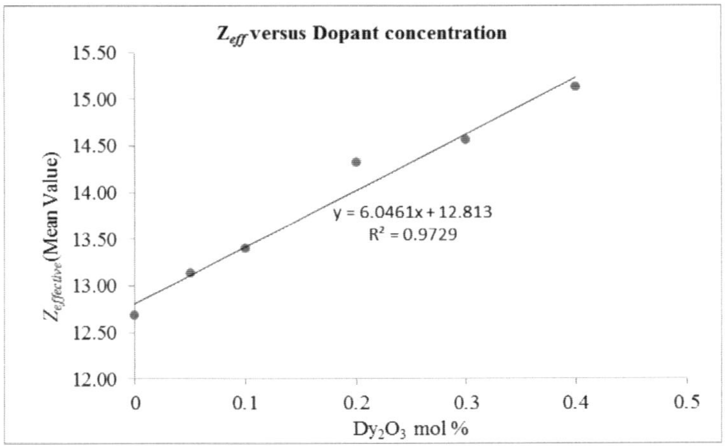

Fig. 11. The effective atomic number of 30CaO-(70-*x*) B_2O_3 glass system (where *x* = 0, 0.1, 0.2, 0.3 and 0.4 mol % of dopant concentrations)

4 Conclusion

The thermoluminescence dosimetric properties, incorporating glow curve structure, linearity, fading, reproducibility and effective atomic number were investigated for 30CaO – (70-x)B_2O_3 : xDy_2O_3 glass system. The optimum concentration for TL response of calcium borate doped with Dy_2O_3 was found to be 0.3 mol%, producing high TL intensity at temperature peak of 286°C. In addition, annealing temperature that offer greater TL intensity and glow peak for the glass sample tested herein was measured to be 300°C, with heating rate of 10°Cs^{-1} .Through the current study, the dosimetry of Dy-doped calcium borate glass have been shown to posses considerable characteristics, including good TL-dose linearity, sensitivity and reproducibility. TL glow curve for studied glass exhibited single, broad curve and has maximum intensity of 286°C. Moreover, 30CaO – (70-x)B_2O_3 : xDy_2O_3 glass offers many other advantages over conventional dosimetry systems, including the calcified tissue equivalence of the dosimetric material. The value of effective atomic number, Z_{eff} is calculated to be 13.56 being close to that of bone. The outcome of the current research is expected to be useful in developing and characterising tailor-made doped borate glass modified with calcium as new potential TLD materials for fraction dose estimation in bone cancer treatment.

Acknowledgement

The authors would like to thank the Ministry of Education (MOE) of Malaysia and Universiti Teknologi Malaysia for financial support through research grant No. 03H28 and the facilities.

References

1. M. Satiago et al *Phys Status Solid A*, **2** 285 (2001)
2. Y. Fukuda, T. Okuno., N. Takeuchi. Thermoluminescence of Borate Glass Containing Copper. Rad. Protectn Dosimentry, **6** 309-312
3. S.S. Rojas., K. Yukimitu, A. S. S. De Camargo., L. A. O. Nunes., & A. C. Hernandes, Undoped and calcium doped borate glass system for thermoluminescent dosimeter. Journal of non-crystalline solids, **32**, 3608-3612 (2006)
4. S.W.S McKeever, Thermoluminescence of Solid, Cambridge University Press (1985)
5. C. K. Sangeeta., S.C. Sabharwal, J. Lumin. **104** 267 (2003).
6. C. K. Sangeeta., S.C. Sabharwal, J. Lumin. **109** 69 (2004).
7. Y. Fukuda, A. Tomita, N. Takeuchi, Phy. Status Solid A **85** 141 (1984).
8. Y. Fukuda, A. Tomita, N. Takeuchi, Phy. Status Solid A **99** 135 (1987).
9. M. Ferhi, K. Horchani-Naifer, and M. Ferid, Combustion synthesis and luminescence properties of $LaPO_4$: Eu (5%). *Journal of Rare Earth*. **27** 182-187 (2009).
10. H. Ohtaki, Y. Fukuda, & N. Takeuchi, Thermoluminescence in calcium phosphate doped with Dy_2O_3. *Radiation Protection Dosimetry*, ***47***, 119-122 (1993).
11. Johnson, P. D., & Williams, F. E. (1950). The interpretation of the dependence of luminescent efficiency on activator concentration. The Journal of Chemical Physics, 18(11), 1477-1483 (1950).
12. Ewles, J., Lee, N,. Studies on the concept of large activator centers in crystal phosphors. J. Electrochem. Soc. 100, 392e398. (1953).
13. M. Prokic, Radiat. Meas. **33** 393 (2001).

14. Mahesh, K., Weng, P. S., & Furetta, C. *Thermoluminescence in Solids and its Applications*. Nuclear Technology Pub. (1989)
15. Gleiter, H. Nanocrystalline materials. In *Advanced Structural and Functional Materials* (pp. 1-37). Springer, Berlin, Heidelberg. (1991)
16. McKeever, S. W.Thermoluminescence of solids (Vol. 3). Cambridge University Press.(1988).
17. A. J. J. Bos, Theory of thermoluminescence. Radiat. Meas. **41**, S45-S56 (2007).
18. A. J. J. Bos, High sensitivity thermoluminescence dosimetry. *Nuclear Instruments and Methods in Physics Research Section B: Beam Interactions with Materials and Atoms*, **1**, 3-28 (2001).
19. G. Busuoli, General characteristics of TL materials. *Applied thermoluminescence dosimetry: lectures of a course held at the Joint Research Centre, Ispra*, **12**, 16-87 (1979).
20. McKinlay, Alastair F. "Thermoluminescence dosimetry." (1981).
21. Hashim, S., Bradley, D. A., Saripan, M. I., Ramli, A. T., & Wagiran, H. The thermoluminescence response of doped SiO 2 optical fibres subjected to fast neutrons. Applied Radiation and Isotopes, 68(4), 700-703 (2009).
22. Furetta, Claudio. "Handbook of Thermo." Luminescence (2003).
23. K. Mizuguchi, & Y. Fukuda, Thermoluminescence and Thermally Stimulated Exoelectron Emission of Ca3 (PO4) 2: Dy and Ca3 (PO4) 2: Tm. Radiat. Prot. Dosim, **84**, 301-305 (1999).
24. Hubbell, J.H. Review of photon interaction cross section data in medical and biological context. Phys Med. Biol. 44, R1-R22 (1999a).

EPJ Web of Conferences **156**, 00003 (2017) DOI: 10.1051/epjconf/201715600003
RCNP 2016

Trace of heavy metals in maternal and umbilical cord blood samples in association with birth outcomes in Baghdad, Iraq

Mayyadah Hasan Rhaif Al-Sahlanee[1,2,*], *Ramzun* Maizan Ramli[1], *Miami* Abdul Hassan Ali[3], *Nada* Fadhil Tawfiq[4], *Nurul* Zahirah Noor Azman[1], *Azhar* Abdul Rahman[1], *Iskandar* Shahrim Mustafa[1], *Nik* Noor Ashikin Nik Abdul Razak[1], *Nor* Zakiah Yahaya[5], *Hana* Mohammed Al-Marri[1], Nur Syuhada Ayob[1] and *Nabela* Zakaria[1]

[1]School of Physics, Universiti Sains Malaysia, 11800 Pulau Pinang, Malaysia
[2]Department of Physiology, College of Medicine, Al-Mustansiriya University, Baghdad, Iraq
[3]Department of Obstetrics & Gynecology, College of Medicine, AL- Mustansiriya University, Baghdad, Iraq
[4]Department of Physics, College of Science, Al-Nahrain University, Baghdad, Iraq
[5]School of Distance Education, Universiti Sains Malaysia, 11800 Pulau Pinang, Malaysia

Abstract. Trace elements are essential nutritional components in humans and inconvenient tissue content that have a significant influence on infant size. The aim of this study is to evaluate the effects of concentration of elements (uranium (U), lead (Pb) and iron (Fe)) and absorption of Pb and Fe on maternal and umbilical cord blood samples. The concentration and absorption of Pb and Fe in blood samples were determined by using atomic absorption spectrophotometry device, while the uranium concentration was determined by using CR-39 detector. Fifty women of age 16–44 years are involved in this study. Results show that the maximum and minimum values of both concentration and absorption in the maternal samples were for Pb and Fe, respectively. In addition, for umbilical cord, the maximum values of concentration and absorption were for Fe and the minimum concentration and absorption were for U and Pb, respectively. A significant correlation between maternal and umbilical cord blood samples was found. This indicates that the Pb, U and Fe elements can easily transfer from maternal to the fetal body which impacts the growth of fetus.

Key words: heavy elements; uranium; lead; iron; maternal; umbilical cord; birth outcomes, Baghdad, Iraq.

* Corresponding author : mayadahhassen73@yahoo.com

1 Introduction

The knowledge obtained from various studies has explain that intrauterine life plays a key role in the evolution of chronic disease in adulthood. The concept suggests that early life conditions can "program" the infants for a spectrum of reverse health outcomes [1-6]. infant growth depends fundamentally on genetic structure, fetoplacental circulation and maternal nutrition[7]. This growth is especially volatile to the impact of environmental risk factors that deactivate the processes during a critical window of susceptibility [8, 9]. The trace elements of Pb, Fe and U are involved in the function of several enzymes and are essential for maintaining health throughout life [10] While, trace elements are fundamental for humans, their particular function are not well understood, especially in the prenatal interval. Even though their amounts are very small in body, they are implicated in various biochemical passageway and are climacteric to the execution of confirmed functions that are necessary to tolerate life [11].

Radiation risks are mostly significant during organogenesis and the early fetal growth period. It induced deformation have a threshold of 100-200 mGy or higher, and are typically associated with central nervous system problems. This happens especially during exposure at 8 to 25 weeks of gestation age [12]. Even at low levels, fetuses are known to be highly volatile to chemical exposure because of their prompt growth, small size, expansion, and restricted purview to dragging mischievous materiality, in comparison with adults [13]. Elements from the maternal can facilely thread into the placenta and transfer to the infants. This makes the infants volatile to metal toxicity [14, 15]. Therefore, even low level mineral exposure early in life, at levels that may not prejudice the pregnant mother, could vandalize fetus growth and development [16]. Furthermore, levels of some metals in blood increase during pregnancy, due to increase in bone rampage [17, 18], and increased of gastrointestinal absorption [19].

Devastation of many military institutions and research center in Iraq after the Gulf war 1991 have caused pervasion of many toxic metals and radioactive source. This has made ordinary people prone to radioactive radiation and toxic elements without their awareness or caution. Millions of Iraqi received higher dose of radioactivity than ordinary background levels which is caused by the pollution of depleted uranium [20], and another significant pollution in the environment is contamination of toxic metals such as lead elements as a result of the human activities and military bombardment [21].

There are numbers of studies published in developing countries that have estimated the effect of trace elements levels on maternal and umbilical cord blood. In Iraq very few studies were published covering population [22-25] In previous studies different concentrations of trace elements such as uranium (U), Lead (Pb) and iron (Fe) have been found in maternal blood and umbilical cord, showing a conflicting cohort with the anthropometric characteristics of the infants [26-35] Several factors may participate to the contradiction in those results from previous studies, such as diversity of the study designs and techniques used, heterogeneity in populations and among others.

The present study is the first report of a cross sectional analysis of heavy elements concentration and absorption as well as uranium concentration in maternal and umbilical cord blood samples at the time of delivery in Baghdad. It aims to determine the presence of correlation between maternal and umbilical cord among pregnant women and the association of lead, uranium and iron with birth outcomes, attending Baghdad maternity hospitals for delivery and correspond potential predictors that may influence both levels.

2 Materials and Method

2.1 Blood Collection

The population for this study involved 50 volunteers (mother – neonate) pairs at accouchement period. The blood samples of maternal and umbilical cord after delivery were collected from all the participants. The maternal blood was collected at the interval of delivery through a venipuncture, and the umbilical cord blood was collected from the placental end without milking the cord.

The blood samples are divided into 2 aliquots. Half of the blood sample were transferred to a EDTA (ethylene diamine tetra acetic acid) heparin containing evacuated tube and used to determine hemoglobin and U concentrations.

The other half of the blood samples were transferred into SSTS (serum separating tubes) and were centrifuged for 10 min to separate the plasma. By using a pipette, 4 ml of plasma was placed in a 7 ml polyethylene tubes for the trace metals analysis. These tubes were frozen for storage and freightage to the laboratories.

Serum separated was used for the estimation of Pb and Fe concentrations and absorption by using atomic absorption spectrophotometry AA-7000 Series. Meanwhile, the U concentration in blood was determine by using CR-39 detector.

2.2 Interview

The interview with each of the respondent was done after delivery. Necessary approval was obtained from them, while information such as health status. Background information were taken, such as health status, nutritional status, intake of Fe during antenatal period, information regarding the previous pregnancies, socio demographic characteristics data (e.g., maternal age, place of living, education, work status, husband work status, tobacco smoking for women and her husband, occupational history, parity of mothers, gestational age which was estimated by using the date for the last menstrual period (LMP), any apparent congenital malformations), information about hemoglobin, blood pressure throughout pregnancy, height and weight. The maternal with hemoglobin <10.0 g/dL were considered as anemic.

2.3 Measurements outcome birth

Birth outcomes includes birth weight, birth length, head circumference, placenta weight, sex, and Apgar scores were measured after delivery. For the purpose of analysis, low birth weight infants are considered for birth weight < 2.500 Kg, and preterm delivery was defined as a birth before 35 completed weeks of GA.

2.4 Irradiation of blood samples

The U concentrations was determined by heating up the blood samples at 37° C for 24 h, using an electric heating incubator type (um100, Memmert, Germany) in order to dry up and oxidize the organic materials. Dried powder blood of 0.5 g was mixed with 0.1 g of methylcellulose ($C_6H_{10}O_5$) which was as a sheath. Then, it was pressed into a pellet which has a dimension of 1 cm diameter and 1.5 mm thickness. The standard blood samples of different U concentration were prepared. Both sides of the pellets were covered with plastic nuclear track detector (CR – 39 Pershore Moulding Ltd, UK) in thickness of 500 μm.

DOI: 10.1051/epjconf/201715600003

In order to cause inherent damage to the CR-39 detector due to ^{235}U (n, f) reaction with the neutron flounce (3.024×10^9 ncm^{-2}), the sample was irradiated by using thermal neutrons from Am – Be neutron source for one week. After irradiation, the CR-39 detectors were etched in 6.25 N of NaOH at temperature of 60 °C, approximately for 5 hr.

The optical microscope (MT 4310H, Meiji, Japan) with magnifications of up to 400x and eye piece of 10x, was used to measure number of track density. It was connected with an optical camera to observed the tracks. The track density (ρ) calculation from eq. (1)

$$\text{Track density } (\rho) = \frac{\text{average of total pits (tracks)}}{\text{area of field view}} \quad (1)$$

2.5 Calculations of Uranium concentrations

The U content in the blood samples was measured by rapprochement between track densities recorded on the CR-39 detectors around the sample pellet and that of the standard samples pellet, from the following relationship [36, 37]:

$$\frac{\text{Cx (sample)}}{\rho\text{x (sample)}} = \frac{\text{Cs (standard)}}{\rho\text{s (standard)}} \quad (2)$$

Where ρx and ρs are the encourage fission track density for unknown sample and standard sample in ($tracks/mm^2$). *Cx* and *Cs* indicate the U concentration for unknown sample and standard sample in (ppm).

2.6 Statistical analysis

The result obtained in this study were analyzed by using Statistical Package for the Social Sciences (SPSS) version 24.0. Descriptive statistics is used to find the frequencies of the values. A Spearman correlation coefficient is used to correlate the maternal and umbilical cord blood of trace elements concentration and absorption. Linear regression model is used to describe the relationship between trace elements and fetal growth. If $p < 0.05$ or < 0.01 it is scrutinized as significant, and if $p \geq 0.05$, it is scrutinized as non-significant.

3 Results and discussion

This population is generally of different socio-economic status and had no receding history of occupational exposure to toxic elements, and also did not take Fe supplementation externally. Table 1 shows the respondents' demographic characteristics. A total of 50 maternal with their infant (25 males and 25 females) were analyzed for current study. These women are healthy and free from any diseases of asthma, hypertension, diabetes, heart problem, Thalassemia, thyroid, gastric, Hepatitis B and C, anemia epilepsy, migraine, renal and TB.

The women involved in this study were aged from 16 to 44 years old. 76% of maternal had high school and higher education, while 24% had less than high school. The percentage of maternal that smoke is 12% during pregnancy. The rates of preterm birth, low birth weight and low Apgar Score were 6%, 14% and 85%, respectively.

The maternal and infant quantitative characteristics at delivery period show in Table 2 the mean value of age is (33.3 ± 7.6 years), body mass index (BMI; weight in kilograms/height in square meters) with mean value (30.7 ± 4.8 (kg/m^2), maternal hemoglobin (9.41 ± 1.80

g/dL), umbilical cord hemoglobin (9.40 ± 1.74 g/dL), systolic and diastolic blood pressure (117.0 ± 10.2, 74.1 ± 6.2 mmHg respectively), pregnancy heart rate and fetal heart rate were (101.0 ± 22.2, 120.0 ± 5.3 bpm respectively). The mean gestational age (GA) at delivery was (38.6 ± 1.3 weeks) and the anthropometric measurements (Birth weight, Birth length, Head circumference and placenta weight) are (2.94±0.93, 39.02±6.8, 35.5±2.3 and 481.05±201.58 respectively).

Table 1. Socio-demographic characteristics

Characteristics			**N**	**percentage**
Maternal	Age	< 30 years old	17	34
		≥ 30 years' old	33	66
	Education	Less than high school	12	24
		High school graduate	22	44
		Higher education	16	32
	Working status	Yes	30	60
		No	20	40
	Husband working status	Yes	44	88
		No	6	12
	Smoking	Yes	6	12
		No	37	74
		Past	7	14
	Parity	0	21	42
		1or more	29	58
Infant	Gender	Male	25	50
		Female	25	50
	Birth Weight	≤ 2.5 kg	7	14
		> 2.5 kg	43	86
	GA at delivery	Pre – Term < 35	3	6
		Early – Term (37-39)	20	40
		Full term ≥ 39	27	54
	5-min Apgar score	< 5	47	85
		≥ 5	3	15

Table 3 and Figure 1 show the elements concentration in maternal and umbilical cord blood samples. The maximum value obtained was 1.27 ppm which represents to iron concentration in umbilical cord blood samples, and the minimum value of concentration is 0.50 ppm for the concentration of iron in maternal blood samples. The concentration of iron in umbilical cord as higher than in maternal blood samples. This indicate that the infants receive iron indirectly from their mother's blood circulation through a prompt and unidirectional passageway.

In maternal the highest levels of elements were recorded in lead level. These higher values could be caused by human activities, industrial wastes resulting from the petroleum, military events during the Gulf wars, and prevailing petrochemical industries in these governorates. Iraq is one of the third-world countries confronting terrifying environmental problems. The use of leaded gasoline, untreated sewage water, motor vehicle exhausts, unchecked wastes of factories, and additionally the remnants of the Gulf wars are considered as the main sources of direct exposure of individuals to lead toxic metals. A Spearman rank test confirmed statistically a positive significant correlation between maternal and umbilical cord blood for Pb, U and Fe elements (r = 0.749, 0.408 and, 0.502 respectively and P < 0.01). This significant positive correlation indicates that even when the mother has a low level of elements, the placenta cannot prevent these elements to transfer from mother to infant.

Table 2. Maternal and infant quantitative characteristics

Characteristics	Mean ± SD	(Min, Max)	Selected percentiles		
			25%	50%	75%
Maternal Age(years)	33.3 ± 7.6	(16, 44)	30	36	39
Body mass index(Kg/m^2)	30.7 ± 4.8	(18.5, 41.1)	27.4	30.6	34.0
Maternal Hemoglobin (g/dL)	9.41 ± 1.80	(7.0, 13.20)	8.00	8.75	10.90
Umbilical Cord Hemoglobin (g/dL)	9.40 ± 1.74	(7.0, 13.20)	8.00	8.90	11.50
Systolic pressure(mmHg)	117.0 ± 10.2	(100, 150)	110	117	120
Diastolic Pressure(mmHg)	74.1 ± 6.2	(60, 90)	70	70	80
Pregnancy heart rate (bpm)	101.0 ± 22.2	(76, 155)	86	93	116
Fetal heart rate(bpm)	120.0 ± 5.3	(120, 144)	130	135	137
GA at delivery (weeks)	38.6 ± 1.3	(35, 41)	38	39	40
Birth Weight (Kg)	2.94 ± 0.93	(2, 4.80)	2.22	2.45	3.93
Birth Length (cm)	39.02 ± 6.8	(30, 53)	34.3	35.9	46.3
Head Circumference (cm)	35.5 ± 2.3	(30, 39.10)	34	35	37.9
Placenta Weight (gm)	481.05 ± 201.58	(300, 895)	320	360	709.3

Table 3. Trace of heavy elements concentration in maternal and cord blood samples.

Elements	Maternal (n= 50)			Umbilical Cord (n = 50)			Statistical Correlations	
	Mean	SD	Min, Max	Mean	SD	Min, Max	r	P - Value
Pb	1.21	0.84	0.009, 2.98	1.18	0.77	0.007, 2.24	0.749	< 0.01*
U	0.95	0.62	0.22, 3.00	0.68	0.39	0.12, 2.51	0.408	< 0.01*
Fe	0.50	0.37	0.06, 1.78	1.27	0.91	0. 04, 4.31	0.502	< 0.01*

Notes: * refers to the significant results.

Table 4 and Figure 2 illustrates the average value of lead and iron absorption in the maternal and umbilical cord blood samples. It is obvious from this table that the mean value of iron absorption is higher than the mean values of other elements. Results shwed statistically significant correlation between maternal and umbilical cord

Table 4. Absorption of heavy elements in maternal and cord blood samples.

Elements	Maternal (n = 50)			Umbilical Cord (n = 50)			Statistical Correlation	
	Mean	SD	Min, Max	Mean	SD	Min, Max	r	P - Value
Pb	0.041	0.006	0.001, 0.034	0.021	0.005	0.0001, 0.037	0.354	< 0.05 *
Fe	0.031	0.030	0.003, 0.729	0.071	0.130	0.002, 1.633	0.440	< 0.05 *

Notes: * refers to the significant results.

Table 5 shows the linear regression models for each anthropometric measurements of the newborns (birth weight, birth length and head circumference) retrograde separately for the maternal and umbilical cord blood trace elements levels. The high value of (B= - 7.78) for Pb concentration in maternal represent that the dependent variable (BL) decreased of (7.78) for every one unit increase in Pb level with significant correlation and CI (-8.97, -6.60). This indicate that any increased of Pb exposure will influence to the anthropometric measurements, and we expected that (BW) increase of (0.3) for every one unit increase in Fe level in maternal for the low value of (B = 0.30). Beta coefficients are used to compare the relative strength of the various prédictions with the model. Pb in maternel for dependent variable (BW) has the larges beta coefficient (-0.91) with a negative signifiant correlation and CI (-1.23, -0.95), Fe in umbilical cord for dependent variable (BL) has the smallest beta (0.25) with a positive non significant correlation and CI (-0.56, 9.67). Thus, one standard deviation decrease in Pb level leads to increase (0.91) in standard deviation of (BW) and one standard deviation increase in Fe level leads to increase (0.25) standard deviation in predicted (BL).

We spotted significant correlation between the Pb, U and Fe level with anthropometric measurements in maternal and umbilical cord blood ($P < 0.05$), while non-significant correlation was found between U, Fe in umbilical cord with birth length ($p > 0.05$). The negative correlation was found between the Pb and U with the anthropometric measurements in maternal and umbilical cord and positive with Fe.

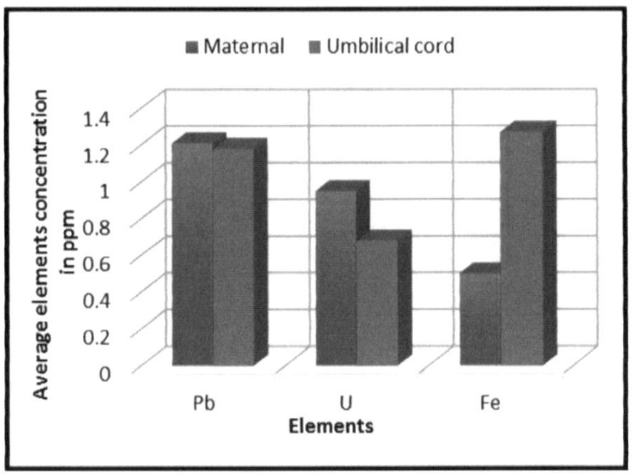

Fig. 1. Trace elements concentrations in blood samples of maternal and umbilical cord

Table 5. Models for adjusted regression coefficients of element levels in maternal and umbilical cord blood for birth outcomes

Position	Dependent Variables	Unstandardized Coefficients		Standardized Coefficients	95% CI	t	P - Value
		B	S.E.	Beta			
(Pb) Maternal	BW	-1.09	0.07	-0.91	(-1.23, -0.95)	-15.13	P < 0.05*
	BL	-7.78	0.59	-0.89	(-8.97, -6.60)	-13.22	P < 0.05*
	HC	-2.14	0.3	-0.72	(-2.74, -1.55)	-7.23	P < 0.05*
(Pb) Umbilical Cord	BW	-0.85	0.10	-0.77	(-1.10, -0.64)	-8.23	P < 0.05*
	BL	-6.12	0.77	-0.76	(-2.98, 0.74)	-7.98	P < 0.05*
	HC	-2.04	0.26	-0.75	(-2.57, -1.52)	-7.80	P < 0.05*
(U) Maternal	BW	-0.58	0.2	-0.37	(-0.95, - 0.15)	-2.28	P < 0.05*
	BL	-3.62	1.48	-0.33	(-6.60, -0.64)	-2.51	P < 0.05*
	HC	-1.61	0.48	-0.42	(-2.51, -0.58)	-3.19	P < 0.05*
(U) Umbilical Cord	BW	-0.70	0.32	-0.31	(-1.38, -0.08)	-2.32	P < 0.05*
	BL	-4.61	2.39	-0.27	(-9.43, 0.20)	-1.90	P > 0.05
	HC	-1.73	0.80	-0.30	(-3.33, -0.10)	-2.11	P < 0.05*
(Fe) Maternal	BW	0.30	0.14	0.30	(0.02, 0.58)	2.18	P < 0.05*
	BL	2.29	1.02	0.31	(0.24, 4.34)	2.24	P < 0.05*
	HC	0.88	0.34	0.35	(0.19, 1.56)	2.58	P < 0.05*
(Fe) Umbilical Cord	BW	0.81	0.34	0.33	(0.13, 1.49)	2.40	P <0.05*
	BL	4.56	2.54	0.25	(-0.56, 9.67)	1.79	P > 0.05
	HC	2.04	0.84	0.33	(0.36, 3.72)	2.45	P < 0.05*

*BW = birth weight, BL = birth length, HC = head circumference

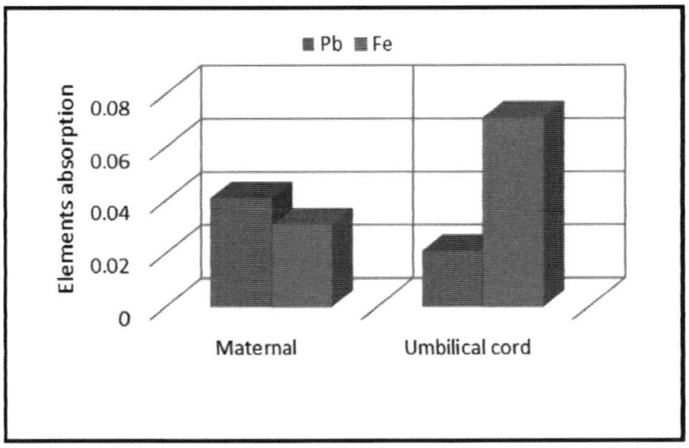

Fig. 2. Absorption of elements in blood samples of maternal and umbilical cord

4 Conclusion

The results evidence the importance of U, Fe and Pb levels in the maternal and umbilical on birth weight. There may be many parameters acting interpret to trigger the condition of low birth weight, and perhaps low birth weight acts merely as a marker of a number of biological lesion. The findings of this study suggest that U and Pb could be specimen of these factors. In Iraq, the present study prepares a base line data on U and Pb concentration and absorption in maternal and umbilical cord samples. Because concentration and absorption of these elements in blood samples are not routinely measured in any health facility. As a result, there are limited data about the propagation and predictors of high blood levels in both general population and high-risk groups including women in childbearing age. Statistically significant positive correlation between maternal and umbilical cord indicates that even when the mother has a low level of elements, placenta cannot prevent these elements to transfer from mother to infant. It was noted that most anthropometric measurements have a négative significant correlation with U, Pb and Fe concentration in maternal and umbilical cord blood.

Acknowledgement

The authors thank to short-term grant of Universiti Sains Malaysia (304/PFIZIK/6312125, 304/PFIZIK/6313249, and 304/PJJAUH/6313152) for funding of this project.

References

1. D. J. Barker, J. Hypertens. **10**, S45 (1992)
2. G. C. Burdge,K. A. Lillycrop, Annu. Rev. Nutr. **30**, 315 (2010)
3. P. D. Gluckman, M. A. Hanson, C. Cooper, K. L. Thornburg, N Engl J Med. **359**, 61 (2008)
4. E. Lurbe, C. Garcia-Vicent, M. I. Torro, F. Aguilar, J. Redon, Hypertension. **63**, 1326 (2014)
5. P. M. Nilsson, E. Lurbe, S. Laurent, J. Hypertens. **26**, 1049 (2008)
6. K. R. Risnes, L. J. Vatten, J. L. Baker, K. Jameson, U. Sovio, E. Kajantie, M. Osler, R. Morley, M. Jokela, R. C. Painter, Int. J. Epidemiol. **40**, 647 (2011)
7. O. Kontic-Vucinic, N. Sulovic, N. Radunovic, Int. J. Fertil. Womens Med. **51**, 116 (2006)
8. P. Grandjean, D. Bellinger, Å. Bergman, S. Cordier, G. Davey-Smith, B. Eskenazi, D. Gee, K. Gray, M. Hanson, P. Van den Hazel, Basic Clin. Pharmacol. Toxicol. **102**, 73 (2008)
9. M. Vrijheid, M. Casas, A. Bergström, A. Carmichael, S. Cordier, M. Eggesbø, E. Eller, M. P. Fantini, M. F. Fernández, A. Fernández-Somoano, Environmental health perspectives. **120**, 29 (2012)
10. B. L. Gulson, K. J. Mizon, M. J. Korsch, J. M. Palmer, J. B. Donnelly, Sci. Total Environ. **303**, 79 (2003)
11. W. Mertz, Science. **213**, 1332 (1981)
12. K. Ohtaki, Y. Kodama, M. Nakano, M. Itoh, A. Awa, J. Cologne, N. Nakamura, Radiat. Res. **161**, 373 (2004)
13. B. Eskenazi, A. Bradman, R. Castorina, Find this article online. (1999)
14. S. Lin, S.-A. Hwang, E. G. Marshall, D. Marion, Am. J. Epidemiol. **148**, 173 (1998)
15. N. Nashashibi, E. Cardamakis, G. Bolbos, V. Tzingounis, Gynecol. Obstet. Invest. **48**, 158 (1999)
16. D. C. Bellinger, Curr. Opin. Pediatr. **20**, 172 (2008)

17. I. Hertz-Picciotto, M. Schramm, M. Watt-Morse, K. Chantala, J. Anderson, J. Osterloh, Find this article online. (2000)
18. E. K. Silbergeld, Environ. Health Perspect. **91**, 63 (1991)
19. T. M. Leazer, Y. Liu, C. D. Klaassen, Toxicol. Appl. Pharmacol. **185**, 18 (2002)
20. S. N. Al-Azzawi, Global research. **1**, 4 (2006)
21. M. Al-Sabbak, S. Sadik Ali, O. Savabi, G. Savabi, S. Dastgiri, M. Savabieasfahani, Bull. Environ. Contam. Toxicol., 1 (2012)
22. I. Al-Saleh, M. A. Khalil, A. Taylor, Arch. Environ. Health. **50**, 66 (1995)
23. B. Kirel, M. Aksit, H. Bulut, Turk J Pediatr. **47**, 125 (2005)
24. R. Iranpour, A. A. Besharati, F. Nasseri, M. Hashemipour, M. Balali-Mood, R. Kelishadi, Saudi. Med. J. **28**, 877 (2007)
25. A. A. Al-Jawadi, Z. W. Al-Mola, R. A. Al-Jomard, BMC Res. Notes. **2**, 47 (2009)
26. H. Guan,F. Piao, Acta Obstet. Gynecol. Scand. Suppl. **90**, 924 (2011)
27. M. Zhu, E. F. Fitzgerald, K. H. Gelberg, S. Lin, C. M. Druschel, Environ. Health Perspect. **118**, 1471 (2010)
28. I. Al-Saleh, N. Shinwari, A. Mashhour, G. E. D. Mohamed, A. Rabah, IInt J Hyg Environ Health. **214**, 79 (2011)
29. X. Xie, G. Ding, C. Cui, L. Chen, Y. Gao, Y. Zhou, R. Shi, Y. Tian, Environ. Sci. Pollut. Res. **175**, 30 (2013)
30. R. Raghunath, R. Tripathi, V. Sastry, T. Krishnamoorthy, Sci. Total Environ. **250**, 135 (2000)
31. E. García-Esquinas, B. Pérez-Gómez, P. Fernández-Navarro, M. A. Fernández, C. De Paz, A. M. Pérez-Meixeira, E. Gil, A. Iriso, J. C. Sanz, J. Astray, BMC Public Health. **13**, 841 (2013)
32. S. Turgut, S. Hacıoğlu, G. Emmungil, G. Turgut, A. Keskin, POL J ENVIRON STUD. **18** (2009)
33. S. Llop, X. Aguinagalde, J. Vioque, J. Ibarluzea, M. Guxens, M. Casas, M. Murcia, M. Ruiz, A. Amurrio, M. Rebagliato, Sci. Total Environ. **409**, 2298 (2011)
34. A. K. M. Tiwari, A. A. Mahdi, F. Zahra, S. Sharma, M. P. S. Negi, Indian J. Clin. Biochem. **27**, 246 (2012)
35. M. H. R. Al-sahlanee, R. M. Ramli, M. A. H. Ali, N. F. Tawfiq, A. A. Rahman, I. S. Mustafa, N. Z. N. Azman, N. N. A. N. A. Razak, N. Z. Yahaya, H. M. Al-Marri, in *Control System, Computing and Engineering (ICCSCE), 2016 6th IEEE International Conference on*. (IEEE, 2016), pp. 360
36. R. L. Fleischer, P. B. Price, R. M. Walker, *Nuclear tracks in solids: principles and applications*. (Univ of California Press, 1975)
37. S. A. Durrani,R. K. Bull, *Solid state nuclear track detection: principles, methods and applications*. (Elsevier, 2013)

Effect of exact Coulomb-exchange calculations on band-head spectra of odd-proton nuclei

Meng-Hock, Koh[1,*] and *Mohamad Nor*, Nurhafiza[1]

[1] *Department of Physics, Faculty of Science, Universiti Teknologi Malaysia, 81310 Skudai, Johor, Malaysia*

Abstract. Previous calculations of band-head energy spectra of odd-mass heavy nuclei in the Hartree-Fock-plus-Bardeen-Cooper-Schrieffer (HF-BCS) framework showed that the agreement with data is better for odd-neutron as compared to odd-proton nuclei. The reason for a poorer agreement with data for the latter have been ascribed to the possible usage of the Slater approximation in calculating the Coulomb-exchange term. In this work, we report the effect of exact Coulomb-exchange calculations on band-head energy spectra of two odd-proton nuclei (namely ^{237}Np and ^{241}Am) as compared to the results obtained using the Slater approximation. We performed self-consistent blocking calculations while taking the breaking of time-reversal symmetry at the mean-field level into account due to the unpaired nucleon. The SkM* and SIII parametrizations of the Skyrme interaction have been employed to approximate the effective nucleon-nucleon interaction while a seniority force is used for the pairing channel. Contrary to what was expected, our preliminary results show no improvement on the band-head spectra as compared to data when the Coulomb-exchange term is calculated exactly.

1 Introduction

Self-consistent blocking calculations of band-head energies of odd-mass nuclei have been recently performed in Ref. [1] within the Hartree-Fock-plus-Bardeen-Cooper-Schrieffer (HF+BCS) framework. Using two choice of Skyrme parametrizations namely the SIII and SkM*, the authors showed that the agreement between the calculated results and experimental data for odd-neutron nuclei (namely ^{235}U and ^{239}Pu) is better than for odd-proton nuclei (namely ^{237}Np and ^{241}Am). A possible reason for the poorer agreement with data for odd-proton nuclei is attributed to the usage of the Slater approximation [2] for the Coulomb-exchange term. As a follow-up to the previous work, we would like to investigate as to what extend the Slater approximation affects the agreement of band-head energies with data. Conversely, we want to check if exact calculations of the Coulomb-exchange term would improve the agreement of the quantity of interest with available data.

While exact calculation of the Coulomb-exchange term is possible, majority of mean-field calculations of various nuclear properties resort to using the Slater approximation due to the former being more cumbersome and requires longer computing time. Some studies on the validity of the Slater approximation have been performed starting from the work of Ref. [3] and more recently in the work of Refs. [4] and [5]. The general consensus was that the Slater approximation constitute a

*e-mail: kmhock@utm.my

good approximation for heavy nucleus while it is less good for light nucleus. Furthermore, the Slater approximation was found to be better in open shell nucleus as compared to nuclei in the vicinity of magic nucleus [5].

However, previous studies involved only even-even (i.e. even with respect to the proton and neutron numbers) mass nuclei whereby only the time-even densities contributes to the Hamiltonian density. In the case of odd-mass nuclei (odd-odd nuclei included), the time-odd densities does not vanish as is the case in even-even nuclei and resulting in the lifting of the Kramers degeneracy of the single-particle levels. The inclusion of the time-reversal symmetry breaking at the mean-field level (due to the time-odd densities) plays an important role in describing for e.g. the quenching of the spin gyromagnetic ratio (see Refs. [6–8]). This preliminary work is therefore, to the best of our knowlege, the first in which exact calculations of the Coulomb-exchange term is performed on top of breaking the time-reversal symmetry at the mean-field level.

2 Details of the approach

This work is based on the HF+BCS framework assuming axial and parity symmetric nuclear shape. The effective nucleon-nucleon interaction is approximated with the phenomenological Skyrme interaction. More specifically, we have make use of the SIII Skyrme parametrization [10]. For the BCS pairing treatment, the seniority force is used whereby the pairing strengths, G_q entering the pairing matrix element, g_q

$$g_q = \frac{G_q}{N_q + 11} \tag{1}$$

has been fitted to the odd-even mass staggering of some actinide nuclei (refer Ref. [1] for details of the fitting procedure). The parameter N_q above refers to the number of nucleons for a given charge q. The retained values are $G_n = 17.15$ MeV and $G_p = 14.00$ MeV for the neutron and proton pairing strengths, respectively. All single-particle states up to 6 MeV above the chemical potential with a smooth cut-off of $\mu = 0.2$ MeV has been considered for the BCS calculations.

In the case of odd-mass nucleus, the lowest-energy solution for a particular blocked K^π configuration is obtained by blocking one single-particle state k nearest to the Fermi level. The third component of the total angular momentum of the blocked state Ω_k is assumed to correspond to the nuclear spin of the whole nucleus (i.e. $\mathrm{K} = \Omega_k$). The blocking procedure was implemented in the code by setting the occupation number v_k^2 of the single-particle state k to 1.

A consequence of blocking a single-particle state is that now the time-odd local densities entering the Hamiltonian densities (defined in terms of the action of the time-reversal symmetry operator) does not vanish as is the case for the ground-state of even-even nucleus. Discussions on these local densities as well as the Hartree-Fock equations can be found for e.g. in Appendix A of Ref. [9].

When performing calculations using the Slater approximation, the contribution from the Coulomb-exchange term to the total binding energy of the nucleus is given by

$$E^{(Slater)}_{exch.Coul.} = -\frac{3}{4}e^2\left(\frac{3}{\pi}\right)^{1/3}\int \mathbf{d^3r}\,\rho_p^{4/3}(\mathbf{r}) \tag{2}$$

The negative sign in the equation (2) reflects the fact that this term provides additional binding energy to the nucleus. On the other hand, the matrix elements of the Coulomb exchange term can be calculated exactly. The details of such procedure are rather lengthy and we would rather refer the readers to the literatures for e.g. Refs. [5, 11, 12]

The HF+BCS solution serves as an intrinsic state solution $\langle\Phi^{\alpha}_{K\pi}|\hat{H}_{eff}|\Phi^{\alpha}_{K\pi}\rangle$ in the Bohr unified model description whereby the total energy of the nucleus $E_{K\pi\alpha}$ is given by [1]

$$E_{K\pi\alpha} = \langle\Phi^{\alpha}_{K\pi}|\hat{H}_{eff}|\Phi^{\alpha}_{K\pi}\rangle - \frac{\langle\hat{\mathbf{J}}^2_{core}\rangle}{2\mathcal{J}} + \frac{\hbar^2}{2\mathcal{J}}\Big[I(I+1) - K(K-1) + \delta_{K,\frac{1}{2}}a(-1)^{I+\frac{1}{2}}\Big(I+\frac{1}{2}\Big)\Big] \quad (3)$$

The Coriolis coupling effect has been neglected except for $K = 1/2$ which is taken care by the decoupling parameter a. The decoupling parameter and the rotational energy $\frac{\langle\hat{\mathbf{J}}^2_{core}\rangle}{2\mathcal{J}}$ have been computed from the microscopic polarized even-even core solution.

Setting the nuclear spin, I, to $I = K$ for band-head, we obtained

$$E_{K\pi\alpha} = \langle\Phi^{\alpha}_{K\pi}|\hat{H}_{eff}|\Phi^{\alpha}_{K\pi}\rangle - \frac{\langle\hat{\mathbf{J}}^2_{core}\rangle}{2\mathcal{J}} + \frac{\hbar^2}{2\mathcal{J}}\Big[2K + \delta_{K,\frac{1}{2}}a(-1)^{I+\frac{1}{2}}\Big(I+\frac{1}{2}\Big)\Big] \quad (4)$$

Using equation (4), the excitation energy $E^{*}_{K\pi\alpha}$ is calculated with respect to the energy of the ground-state band. The moment of inertia $\mathcal{J}$ has been calculated by using the Inglis-Belyaev formula (see e.g. Ref. [14]) omitting the blocked single-particle state and its conjugate state.

3 Results

Calculations of band-head energies have been performed for ^{237}Np and ^{241}Am using both the Slater approximation and exact calculations of the Coulomb-exchange term. The results are plotted in Fig. 1 and Fig. 2 and compared to experimental data. The excitation energies are tabulated in Table 1 together with the calculated values for the moment of inertia. The latter is given in terms of energy parameter $A = \frac{\hbar^2}{2\alpha\mathcal{J}}$ and was obtained after multiplying the moment of inertia by a factor of 1.32 as proposed by Ref. [13]

From the figures, it appears that spectra obtained from exact Coulomb-exchange calculations and Slater approximation are almost similar. Only some blocked configurations obtained from exact calculations are pushed further apart, resulting in a less compress spectra. However, the root-mean-square energy deviations of 562 keV and 521 keV for exact calculations and Slater approximation, respectively, show that the exact calculations do not improve the agreement with data as was initially thought.

4 Conclusion

We have performed exact Coulomb-exchange calculations for band-head energies of two odd-proton nuclei and to compare the results with those obtained using the usual Slater approximation. The SIII Skyrme parametrization and the seniority force were employed for this study. It was found that the energy spectra obtained from both sets of calculations to be rather similar with no significant improvement in the agreement to data when exact calculations were performed. Nevertheless, it is still premature to conclude that exact Coulomb-exchange calculations has no meaningful impact when band-head energies are concerned. Further work is needed especially by focusing on regions whereby more experimental data are available for comparison.

5 Acknowledgement

M.H.K would like to acknowledge Dr. Yap Yung Szen for the financial support rendered to him through the Universiti Teknologi Malaysia Encouragement Grants (Vot. No. 11J23). The author would also like to extend his gratitude to Prof. Philippe Quentin (CENBG) for a fruitful discussion on this particular work.

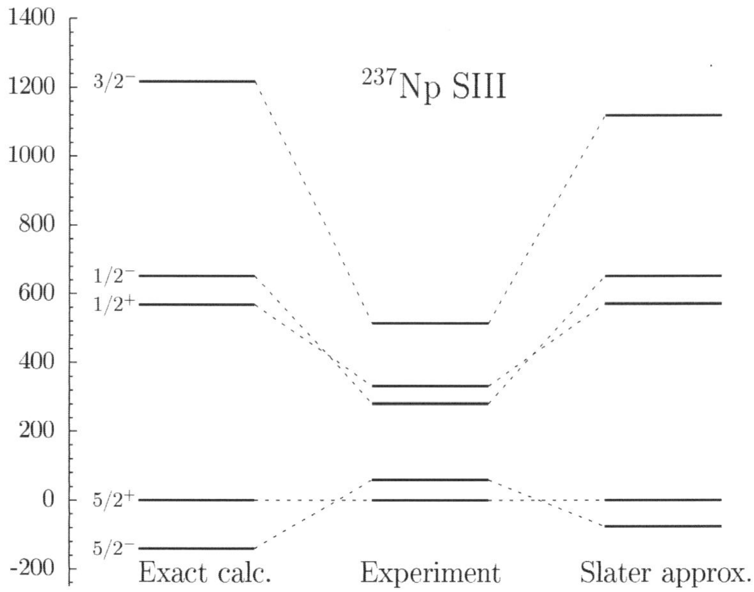

Figure 1: Comparison of the band-head energy spectra of ^{237}Np obtained using the two different approaches for the calculations of the Coulomb-exchange term as compared to experimental data [17]. The results pertaining to Slater approximation and exact calculations are explicitly indicated.

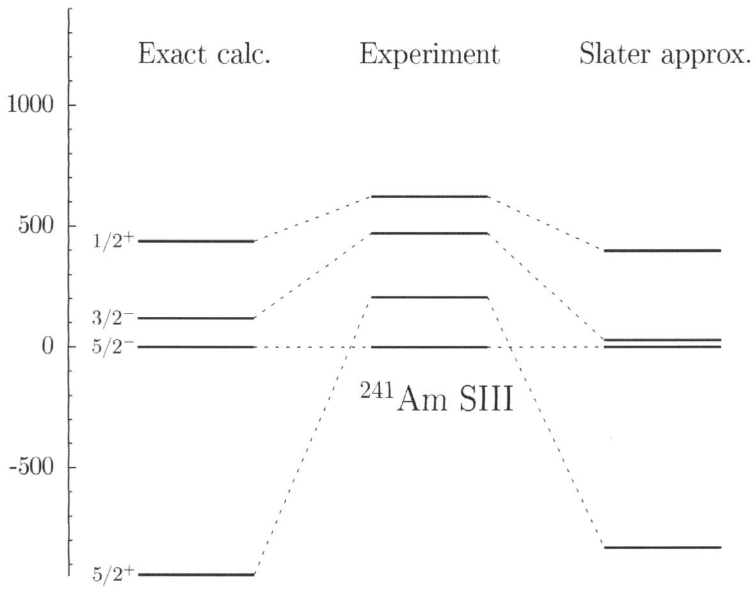

Figure 2: Similar to Fig. 1 but for ^{241}Am nucleus.

Table 1: Excitation energies of band-head states $E^*_{K\pi\alpha}$ and energy parameter ($A = \frac{\hbar^2}{2\beta\mathcal{J}}$) obtained from exact Coulomb-exchange calculations as compared to when Slater approximation is being used. The constant $\beta = 1.32$ of Ref. [13] is to account for the Thouless-Valatin correction term. Experimental data taken from Ref. [17] are included for comparison.

Nucleus	K^π	$E^*_{K\pi}$ [MeV]			A [keV]		
		Exact	Slater	Exp.	Exact	Slater	Exp.
^{237}Np	$5/2^+$	0	0	0	10.62	10.85	4.7
	$5/2^-$	-140	-76	59.5	8.56	9.27	6.2
	$1/2^-$	652	652	281.35	9.08	9.90	6.9
	$1/2^+$	568	571	332	8.16	8.82	6.2
	$3/2^-$	1218	1120	514.2	9.65	10.19	6.3
^{241}Am	$5/2^-$	0	0	0	8.32	9.02	5.9
	$5/2^+$	-943	-829	205.9	10.32	10.59	4.2
	$3/2^-$	119	28	471.8	8.90	9.52	6.5
	$1/2^+$	437	398	623.1	7.76	8.44	6.0

References

[1] Meng-Hock Koh, Dao Duy Duc, T.V. Nhan Hao, Ha Thuy Long, P. Quentin and L. Bonneau, Eur. Phys. J. A **52**, 3 (2016)
[2] J. C. Slater, Phys. Rev. **81**, 385 (1951)
[3] C. Titin-Schnaider and P. Quentin, Phys. Lett. B **49**, 397 (1974).
[4] J. Skalski, Phys. Rev. C **63**, 024312 (2001)
[5] J. Le Bloas, Meng-Hock Koh, P. Quentin, L. Bonneau, and J. I. A. Ithnin, Phys. Rev. C **84**, 014310 (2011)
[6] P. Quentin, L. Bonneau, N. Minkov and D. Samsœn, Int. J. Mod. Phys. E **19**, 611 (2010)
[7] L. Bonneau, J. L. Bloas and P. Quentin, Int. J. Mod. Phys. E **20**, 252 (2011)
[8] L. Bonneau, N. Minkov, Dao Duy Duc, P. Quentin, and J. Bartel, Phys. Rev. C **91**, 054307 (2015)
[9] Meng-Hock Koh, L. Bonneau, P. Quentin, T. V. Nhan Hao, and H. Wagiran, Phys. Rev. C **95**, 014315 (2017)
[10] M. Beiner, H. Flocard, Nguyen Van Giai and P. Quentin, Nucl. Phys. A **238**, 29 (1975)
[11] D. Vautherin, Phys. Rev. C **7**, 296 (1973)
[12] P. Quentin, J. Phys. (France) **33**, 457 (1972)
[13] J. Libert and M. Girod, J.-P. Delaroche, Phys. Rev. C **60**, 054301 (1999)
[14] D. W. L. Sprung, S. G. Lie, M. Vallieres and P. Quentin, Nucl. Phys. A **236**, 37 (1979)
[15] K. J. Pototzky, J. Erler, P.-G. Reinhard and V. O. Nesterenko, Eur. Phys. J. A **46**, 299 (2010)
[16] H. Flocard, P. Quentin, A. K. Kerman and D. Vautherin, Nucl. Phys. A **2013**, 433 (1973)
[17] A. K. Jain, R. K. Sheline, P. C. Sood and Kiran Jain, Rev. Mod. Phys. **62**, 2 (1990)

 DOI: 10.1051/epjconf/201715600005

Statistical Neutron Emission Model for Neutrino Nuclear Response

I.H.Hashim[1], H.Ejiri[2], F.Othman[1], S.S.Saroni[1], W.N.Amelia[1], S.A.Hamzah[1] and K.C.Wei[3]

[1]Dept. of Physics, Faculty of Science, Universiti Teknologi Malaysia (UTM), 81310 Johor Bahru, Malaysia

[2]Research Center of Nuclear Physics (RCNP), Osaka University, Ibaraki-City, 560-0053 Osaka, Japan

[3]Faculty of Nuclear Engineering, Universiti Teknologi Malaysia, 81310 Johor Bahru, Malaysia

Abstract. The studies of neutrino fundamental properties are widely investigated by double beta decay and inverse beta decay. Recent muon capture experiment provides promising ways to directly evaluate the neutrino nuclear responses. This study provides theoretical explanation for neutrino nuclear responses by muon capture experiment. The effects of muon binding energy towards nuclear excitation region and pre-equilibrium (PEQ) and equilibrium (EQ) neutron decay mode will be discussed. The interpretation of muon strength model by comparison of calculator output and recent experimental data may provide relevant information towards determination of nuclear matrix element (NME) and its missing parameter.

Keywords: Neutron Emission Model, Neutrino Nuclear Response, Muon Charge Exchange Reaction, Nuclear Matrix Element, Muon Capture Strength.

1 Introduction

Theoretical calculations of the weak responses (matrix elements) are challenging since they are sensitive to the nuclear correlations and nuclear medium effects. Thus, experimental work and related theoretical verification are very valuable. Recently nuclear responses of $M(\beta^-)$ have been studied extensively by using high-energy resolution $(^3He, t)$ experiments at RCNP. On the other hand, observation of $M(\beta^+)$ are hard as of no high energy resolution nuclear probes.

It has been pointed out in early 1970s that muon capture reactions (MCR) are alternative to probe the β^+ responses [1]. The β^+ strengths are obtained by measuring prompt [6] and delayed gamma rays [9-11] from (μ, xn) reactions with x being the number of the emitted neutrons. So far nuclear gamma rays from MCR were measured to study the nuclear reaction mechanisms [3,4]. The prompt gamma rays originating from bound states are excited by μ capture help the probing of β^+ responses at low lying bound states [6,9]. It is hard to extract the β^+ strength for individual states.

From the Quasi Random Particle Approximation (QRPA) model, the partial capture rate of low lying excited states contributes to larger error to the weak response of these reactions

[7,8]. Furthermore, the complicated data analysis of $M(\beta^+)$ responses *from* $(t,{}^3He)$ reaction acknowledges that MCR is better candidate in investigating the higher excited states properties. In comparison to QRPA model, this statistical modelling might be able to simulate the process of neutron decay assisting the MCR. It was the first model so study the muon strength distribution contributing to the double beta decay β^+ side information [9].

The present study aims to develop the theory to explain the neutrino nuclear response from observation of neutron emission following MCR. The parameters affecting the neutron emission are identified. The neutron spectrum from the neutron emission simulator will be compared with the previous data on muon capture experiment [3]. The relationships of each parameter are identified and express as a function of atomic mass number. These expressions will be implemented to produce the expected isotope mass distribution for muon capture experiment. The μ capture strength distribution as a function of the excitation energy is derived so as to reproduce the measured reaction branches on the basis of the equilibrium (EQ) and PEQ neutron emission mechanism.

The results of this investigation will shed light on the unknown effective axial-vector coupling constant, $g_{A,eff}$ and spin iso-spin nuclear parameter, g_p which lead to big uncertainties in determining the nuclear matrix element. It will also help the theorists group to decide whether these unknown parameters are suitable with their model or requires adjustment to their parameters.

2 Theory of Neutron Statistical Model

A useful model to predict the muon captures strength by implementing the normal distribution for neutron emission events. By using the basis of C and C++ program, the statistical simulation generates the neutron emission spectrum and compares with the available experimental data. To understand the relationship of neutron emission with the weak nuclear response, the model is constructed into three main parts; the binding energy calculator, neutron emission simulator and isotope population calculator.

The binding energy calculator uses the realistic neutron binding energy and muon binding energy from the raw nuclear mass data reported in nuclear database. This information is important in the neutron emission simulator and isotope population calculator. The fast neutron or pre-equilibrium (PEQ) and slow neutron emission or equilibrium (EQ) are studied by using the neutron emission simulator. The neutron energy spectrum are compared with the previously reported neutron spectrum from muon capture experiment to get the optimize parameters. The combination of best optimize parameter value will be implement in the simulation of isotope population and compare with the experimental data of ^{100}Mo [9].

Fig. 1 illustrates the layout of the present muon charge exchange reaction (MCER). The muon captures on ^{100}Ru excites the nuclei to some bound states at unknown excitation levels. Observation by experiments provides mass distribution from gamma and muonic x-rays related to beta decay process. By studying the mode of neutron decay, the nuclear excitation range at initial bound state after muon captures can be obtained. Comparison between the model isotope population output and the experimental data may help confirm the nuclear excitation region.

 DOI: 10.1051/epjconf/201715600005

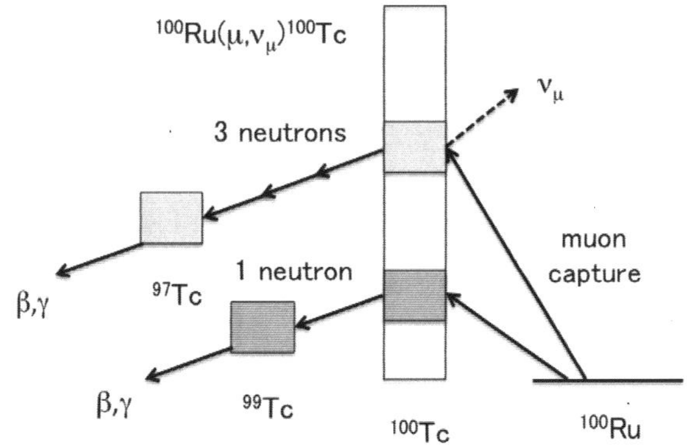

Fig 1: Present MCER ^{100}Ru(μ, ν_μ, xnγ)$^{100-x}$Tc ν⁻ τ (β)+ responses

In order to reduce the ambiguity of the calculation, several assumptions are made. The excitation energy, nuclear temperature and atomic mass number often influenced the level density parameter value. Previous studies shown several approximations and models are made for the systematic studies to estimate the accurate value of level density parameter [12]. In this model, we assume that for ^{100}Mo target the level density parameter are set to constant at 14 MeV.

The probability of proton emission was expected to be very small where the probability of proton emission begins after three or more neutrons emission. In previous study by [13], shown that most MCR on enriched target shows about 90% emission from neutron through beta decay. This might cause by the Coulomb force by the proton, which cause higher potential barrier in removing one proton compare to one neutron.

The neutron emission decay mode pertained as challenging issue since it is difficult to differentiate the neutron decay process. In order to study the influence of PEQ neutron emission, the nuclear temperature are set to be in the factor of 4. The value is assumed from the observation from previously observed neutron emission spectrum. It was expected that this factor are bigger for heavy nuclei.

In muon irradiation experiment, most muon will be captured and some of them will decay. The MCR may excite the nuclei up to 100 MeV due to its high momentum transfer [9][11]. The captured muons are bound to the nuclei at some time interval before muon neutrino releases. The energy to bind the muon with the nuclei are large correspond to the K-shell energy of the atom. Thus, the heavy nuclei possess higher muon binding energy compare to light nuclei. After the bound muon escapes and emitted muon neutrino, the nuclear excitation energy immediately decreases by the factor equals to the muon binding energy. It was expected that at excitation energy of about 21 MeV, about two neutrons are emitted. If the excitation energy is greater than the neutron binding energy, more neutron emission are favorable.

The compound and pre-compound nuclei will be formed associates with the emission of EQ and PEQ neutron. The neutron emission spectrum by the Time-of-Flight (TOF) measurement [4][13], shows that the maximum neutron energy are less than 30 MeV and there are small percentage of slow neutrons emitted at higher than 30 MeV range. The ratio of fast neutron to slow neutron are unknown but it was expected to be in the range of 12% to 30%. By assuming that the probability of neutron emission is equation (1.0) where the first term represent the slow neutron emission which emitted when the nuclear temperature are achieved and the second term represents the fast neutron before achieving nuclear temperature. The ratio of neutron emission during EQ and PEQ are studied as a function of atomic mass number, A.

$$S(E_n) = a\ \exp(-E_n / T_{EQ}) + b\ \exp(-E_n / T_{PEQ}) \tag{1.0}$$

Where $S(E_n)$ is the neuton emission distribution, a and b are constant and the ratio of b/a should be kept between 15% to 30% [9]. The T_{EQ} represents the nuclear temperature at EQ conditions while T_{PEQ} is nuclear temperature at PEQ conditions. The nuclear temperature of slow and fast neutron also differs in some factors as reported in [3]. Since the nuclear temperature factors for ^{100}Mo have not been reported, we use the comparison method to get the optimize value of the factor.

3 Results and Discussion

The ratio of b/a from equation (1.0) is manipulated. Three conditions are considered to study their reproducibility towards experimental results. These comparisons might provide inaccurate estimation of neutron energy as well as the neutrino responses. Fig. 2 shows the neutron emission spectrum at b/a ratio of 0.1% to 1%. For b/a ratio of 0.1%, PEQ neutron emissions are about 30% to 100%. While the ratio of 0.6% and 1%, the PEQ neutron emissions increses from 20% to 70% and 2% to 15% respectively. The values increases as the atomic mass number increases. The higher b/a ratio provides more PEQ neutron emissions.

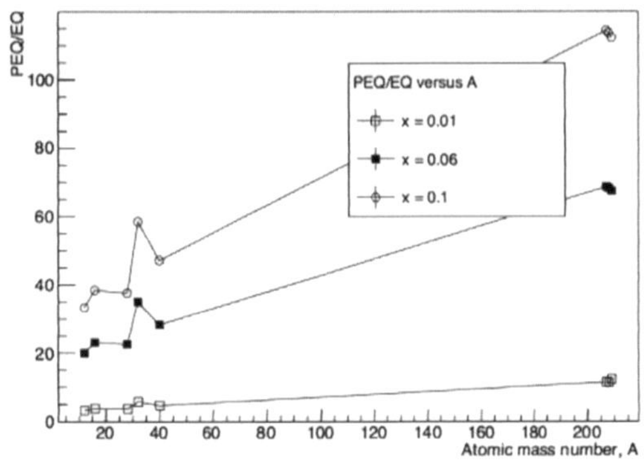

Fig. 2: Ratio of PEQ and EQ neutron emission event at different b/a ratio (x =0.01, 0.06 and 0.1)

For ^{100}Mo target, the muon binding energy are expected to be 4 MeV. If the muon excites the nuclei at 100 MeV and the probabilty of one neutron to emit is 11 - 14 MeV, the maximum excitation energy after the muon neutrino are release at the range of 30 to 40 MeV. Fig. 3 investigate the PEQ neutron emissions when the maximum excitation energy differs. The overall PEQ neutron emissions are in the range of 5% to 100% increases as a function of atomic mass number.

 DOI: 10.1051/epjconf/201715600005

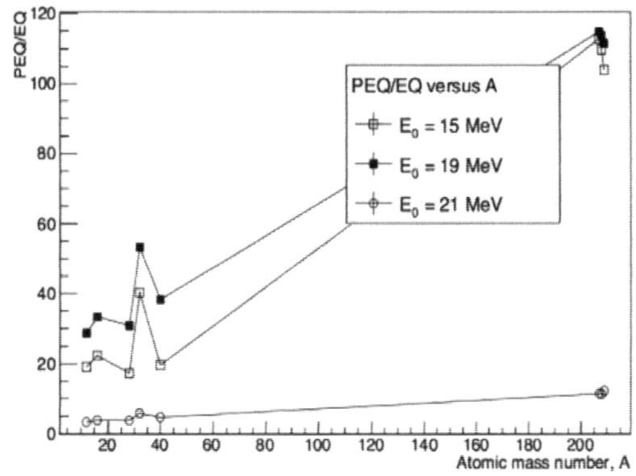

Fig. 3: Ratio of PEQ and EQ neutron emission event at different maximum excitation energy, E_0

Fig. 4(a) shows the comparison of experimental data by Evseev (1972) [4] with the model using b/a ratio of 0.1% and excitation energy of 21 MeV. The χ^2 analysis are used to check which cases are the nearest to experimental data point. All points have very low value of $\Sigma\chi^2$, however the combination shows in Fig. 4(a) are the optimize parameters. Since the ^{207}Pb are heavy nuclei, we would expected that ^{100}Mo shall have almost similar parameters.

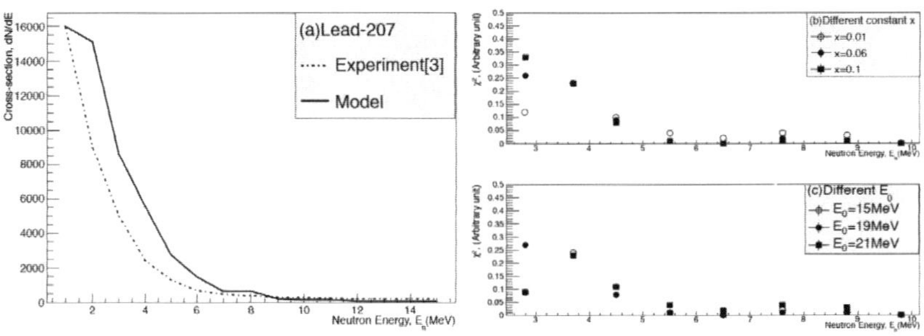

Fig. 4: (a) Comparison for ^{207}Pb with the model (b) χ^2 analysis for different b/a ratio and (c) χ^2 analysis for different excitation energy

By using these parameters, the strength distribution of muon capture will be deduced. The muon capture on enriched Mo produces 5 radioisotopes coming from zero to five neutron emission channel [9]. In order to reproduce this experimental data, the capture strength distribution are sampled from 0 to 70 MeV as shown in Fig. 5. The peak strength was appointed in the range of 11 MeV to 14 MeV due to spin-isospin correlation. Most captures at low excitation energy was followed by one neutron emission. A very low population of isotope detected at higher neutron emission.

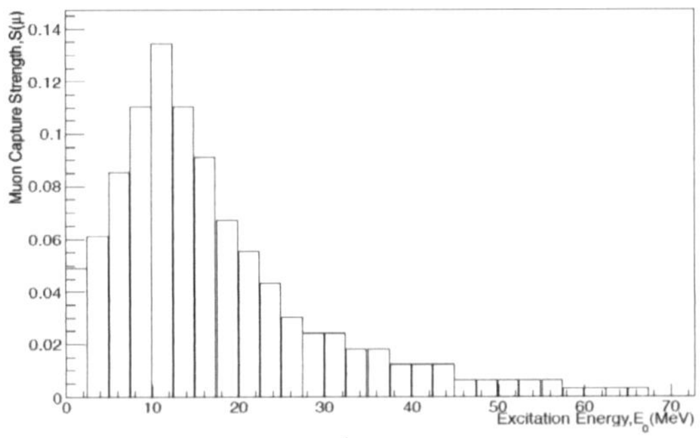

Fig. 5: The capture strength model for comparison with experimental data of ^{100}Mo

From Fig. 6, the combination of optimize parameter presents similar observation with the experimental data. Most captures at low excitation energy was followed by 1 neutron emission. A very low population of isotope detected at higher neutron emission. However, there are some point where the experimental observation is two times compare to the model. This only happens at nuclei involved in more than three neutron emission. This is due to the factor of nuclear temperature which are sets to be in the factor of 4. Systematic study of all the parameters are necessary so that the optimize value for all isotope range can be studied.

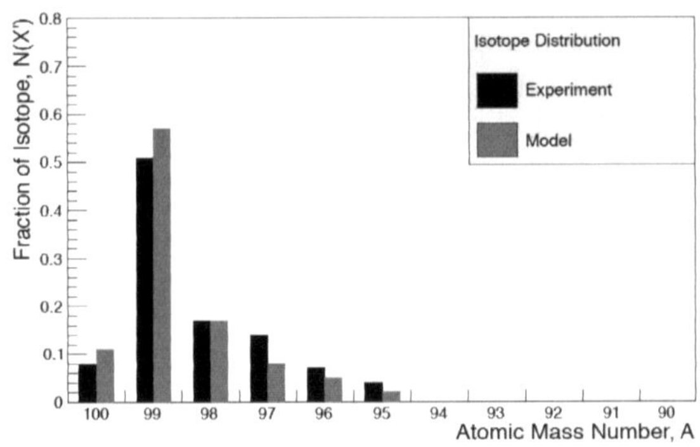

Fig. 6: The isotope population comparison for ^{100}Mo(μ,xnγ)$^{100-n}$Mo

4 Conclusions

Muons are a useful probe to study the nuclear structure information that is relevant to $\beta\beta$ decay experiments. It can provide informative discussion on the higher excited states of a nuclei by using weak probe, which are similar to the responses observed in β^+ side of $\beta\beta$ decay. The statistical modeling proposed that the initial capture strength accumulates at peak around 11 to 14 MeV. The model successfully reproduced the experimental observation using enriched ^{100}Mo target. The theoretical calculations for the determination of NME by QRPA are necessary for MCR.

Author would like to express her gratitude to Universiti Teknologi Malaysia Research Grant VOT Q.J130000.2726.11J23 for the funding of this research.

References

1. H. Ejiri, *Proc. EM interactions*, Sendai. (1972).
2. H. Ejiri, *Phys. Rep. C* **338** 265 and refs. therein (2000).
3. D.F. Measday, *Physics Reports* **354**, 243 and refs. therein (2001).
4. D.F. Measday, et al., *Phys. Rev.C* **73**, 045501. (2006).
5. J. Suhonen, et al., *Czech. J. Physics* **56** 519. (2006).
6. V. Egorov, et al., *Czech. J. Phy.* **56** 523. (2006)
7. H. Ejiri, I.H. Hashim, D. Freckers, private communication. (2014).
8. I.H. Hashim, T.H. Kosmas. private communication. (2014).
9. I.H. Hashim, *PhD thesis.* Osaka University. (2015).
10. I.H. Hashim, *Meson 42th Spring edition*. Japan. (2015).
11. H. Ejiri et. al. *Journal of Physical Society Japan*, **82** 044202. (2013).
12. R. Capote. *Nuclear Data Sheets* **110** 3107–3214. (2009).
13. P. Singer. *Muon Physics Conference*. Colorado State University Fort Collins, Colorado, 39-87. (1971).

DOI: 10.1051/epjconf/201715600006

Terrestrial Gamma Radiation Dose Rate of West Sarawak

A. Izham, *A.T.* Ramli, *W.M. Saridan* Wan Hassan, *H.N.* Idris, *N.A.* Basri

Science Faculty, Universiti Teknologi Malaysia, Skudai, 81310. Johor Bahru, Malaysia

Abstract. A study of terrestrial gamma radiation (TGR) dose rate was conducted in west of Sarawak, covering Kuching, Samarahan, Serian, Sri Aman, and Betong divisions to construct a baseline TGR dose rate level data of the areas. The total area covered was 20,259.2 km^2, where in-situ measurements of TGR dose rate were taken using NaI(Tl) scintillation detector Ludlum 19 micro R meter NaI(Tl) approximately 1 meter above ground level. Twenty-nine soil samples were taken across the 5 divisions covering 26 pairings of 9 geological formations and 7 soil types. A hyper-pure Germanium detector was then used to find the samples' ^{238}U, ^{232}Th, and ^{40}K radionuclides concentrations producing a correction factor $C_f = 0.544$. A total of 239 measured data were corrected with C_f resulting in a mean Dm of 47 ± 1 nGy h^{-1}, with a range between 5 nGy h^{-1} – 103 nGy h^{-1}. A multiple regression analysis was conducted between geological means and soil types means against the corrected TGR dose rate Dm, generating $D_{G,S} = 0.847D_G + 0.637D_S - 22.313$ prediction model with a normalized Beta equation of $D_{G,S} = 0.605D_G + 0.395D_S$. The model has an 84.6% acceptance of Whitney-Mann test null hypothesis when tested against the corrected TGR dose rates.

1 Introduction

Humans are bombarded with natural gamma radiation daily (IAEA, 1989), originating from solar radiation and terrestrial gamma radiation of geological background and soil (UNSCEAR, 2000).

Radionuclides that have long half-lives that are formed during the formation of stars make up trace parts of today's geological background (Valkovic, 2000) contributing to varying levels of gamma radiation. These are naturally occurring radionuclides called primordial radionuclides. Although they exist in trace amounts in nature, the effects of gamma radiation over time may add up to cause stochastic effects (UNSCEAR, 2000).

This study focuses on terrestrial gamma radiation of west Sarawak, covering 5 divisions, Kuching, Samarahan, Serian, Sri Aman, and Betong, and explores the relationship between gamma radiation dose rates with geological background and soil types of the areas.

DOI: 10.1051/epjconf/201715600006

2 Research Methodology

Data Collection

A total of 241 in-situ measurements were made in the study areas covering a total of 20,259.2 km^2:

- Kuching (4,559.5 km^2)
- Samarahan (2,927.56 km^2)
- Serian (2,405 km^2)
- Sri Aman (6186.4 km^2)
- Betong (4,180.74 km^2)

(Laman web rasmi Pentadbiran Bahagian Kuching, Pentadbiran Bahagian Samarahan, Majlis Daerah Serian, Pentadbiran Bahagian Sri Aman, Majlis Daerah Sri Aman, Jabatan Pertanian Sarawak, 2017).

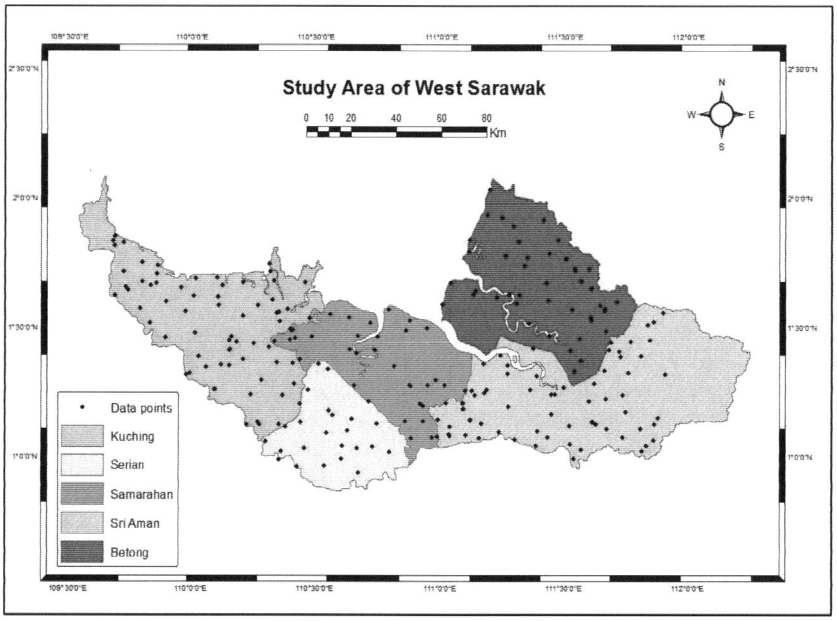

Fig 1. Study areas of west Sarawak, Kuching, Serian, Samarahan, Sri Aman, and Betong divisions and 241 terrestrial gamma radiation dose rate measurement locations (data points)

The dose rate readings were taken using a gamma radiation scintillation detector Ludlum 19 μR h^{-1} (micro Roentgen per hour). Sarawak map, (Directorate of National Mapping Malaysia, 2010), Sarawak geology map (Minerals and Geoscience Department Malaysia, 2013) and Sarawak soil map (Jabatan Pertanian Kuching Sarawak, 2008) were used to aid with random sampling method in consideration of geological background, soil distribution, and ease of access for data points selection.

Correction Factor

Twenty-nine soil samples were taken throughout the 5 divisions and a Hyper-Pure Germanium detector at Science Faculty Universiti Teknologi Malaysia was used to examine their ^{238}U, ^{232}Th, and ^{40}K radionuclides content. The radionuclides concentration values were used to calculate terrestrial gamma radiation dose rates of the soil samples using the following equation (UNSCEAR, 2000):

$$D_C = 0.462C_{238\text{-}U} + 0.604C_{232\text{-}Th} + 0.042C_{40\text{-}K} \quad (2.1)$$

where D_C is calculated gamma radiation dose rate, $C_{238\text{-}U}$ is ^{238}U radionuclide concentration, $C_{232\text{-}Th}$ is ^{232}Th radionuclide concentration, and $C_{40\text{-}K}$ is ^{40}K radionuclide concentration.

Calculated dose rates D_C for the 29 samples were then compared to their respective in-situ dose rate readings by linear regression to obtain a correction factor C_f to be applied to all in-situ measurements producing corrected dose rates Dm (Sanusi, 2014).

Prediction Model

The relationship between geological formations and soil types of the study areas to the dose rates was explored. There are 9 geological formations and 7 soil types present in west Sarawak, with a total of 26 geology-soil pairings. The geological formations and soil types present in west Sarawak are as the following (Directorate of National Mapping Malaysia, 2010) :

Geological formations :

1. Igneous (G1)
2. Quaternary (G2)
3. Tertiary (G3)
4. Late Cretaceous – Miocene (G4)
5. Late Cretaceous (G5)
6. Late Jurassic – Late Cretaceous (G6)
7. Jurassic – Cretaceous (G7)
8. Triassic (G8)
9. Pre-Triassic / pre-Late Carboniferous (G9)

Soil types :

1. Alluvial and arenaceous (S1)
2. Gley (S2)
3. Podzolic (S3)
4. Organic (S4)
5. Skeletal (S5)
6. Podzol (S6)
7. Thionic (S7)

Mean dose rates for geological formations and soil types of west Sarawak were compiled and compared to dose rate readings of their respective corrected dose rates Dm. Multiple regression analysis was then conducted between corrected dose rates Dm, geological formations dose rates D_G, and soil types dose rates D_S to find their respective contribution to the Dm dose rates (Tzortzis, 2002; Muneer, 2013; Sanusi, 2014). The resulting mathematical

 DOI: 10.1051/epjconf/201715600006

model was then applied to each data points and tested for statistical significance against D*m* using Whitney-Mann test.

3 Results and Analysis

Data

To obtain data normality, 2 outliers were excluded from the 241 collected dose rate readings. The calculated dose rate D_C from equation (2.1) of the 29 samples were then compared to their respective in-situ dose rate readings, producing a correction factor $C_f = 0.544$ from a linear fitting of $y = 0.544x + 8.13$. This correction factor C_f was then applied to the remaining 239 dose rate readings. Mean corrected dose rate reading D*m* is 47 ± 1 nGy h^{-1} with a range between 5 nGy h^{-1} – 103 nGy h^{-1}.

The distribution is deemed normal due to its acceptable histogram trendline from Figure 2, skewness of 0.469, and kurtosis of 0.263.

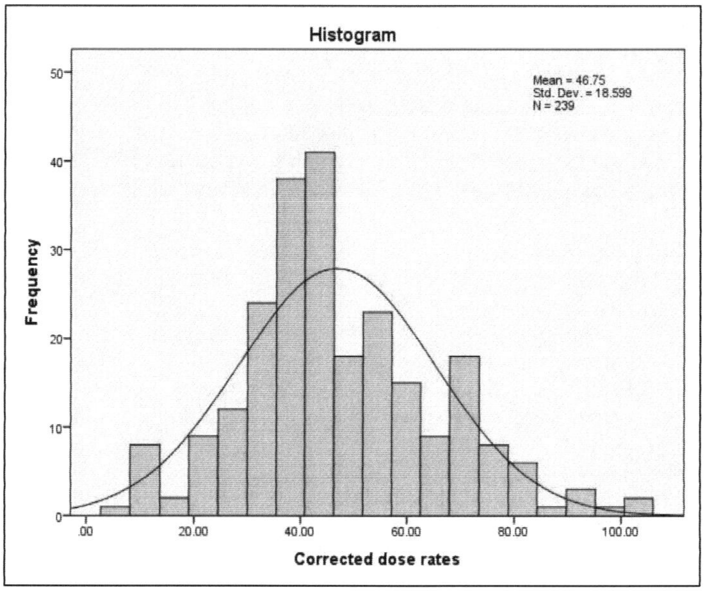

Fig 2. Histogram of corrected dose rates with trendline

Prediction model

A multiple regression of geological formation dose rates, soil types dose rates, and corrected dose rates generates the equation:

$$D_{G.S} = 0.847D_G + 0.637D_S - 22.313 \quad (3.1)$$

The normalized Beta coefficient equation is:

$$D_{G.S} = 0.605D_G + 0.395D_S \quad (3.2)$$

This prediction model (3.1) is applied to all 26 pairs of geology-soil pairings and the comparison of geology-soil pairing predicted dose rates to the corrected dose rates D*m* and Whitney-Mann test of the comparisons are as in Table 1.

Table 1. Comparisons of geology-soil pairing predicted dose rates ($D_G \cap D_S$) to corrected dose rates D*m* and their respective Whitney-Mann test p-values.

D_G	D_S	Corrected dose D*m*	Predicted $D_G \cap D_S$	Whitney-Mann test p-value	Null test decision
G1	S3	40	48	0.013	rejected
G1	S5	57	55	0.367	accepted
G2	S2	40	40	0.519	accepted
G2	S3	40	41	0.224	accepted
G2	S4	37	36	0.113	accepted
G2	S5	60	49	1.000	accepted
G2	S7	39	36	0.699	accepted
G3	S2	37	45	0.102	accepted
G3	S3	46	46	0.071	accepted
G3	S4	30	41	0.487	accepted
G3	S5	50	54	0.039	rejected
G4	S3	34	38	0.163	accepted
G4	S5	43	47	0.577	accepted
G5	S2	60	61	1.000	accepted
G5	S3	62	62	0.581	accepted
G5	S4	49	57	0.102	accepted
G5	S5	71	70	0.222	accepted
G6	S3	39	40	0.247	accepted
G6	S5	35	48	0.037	rejected
G7	S3	41	47	0.033	rejected
G7	S5	51	55	0.102	accepted
G8	S2	57	54	1.000	accepted
G8	S3	56	54	1.000	accepted
G8	S4	45	49	1.000	accepted
G8	S5	64	63	0.487	accepted
G9	S3	44	45	0.094	accepted

From a total of 26 pairings, 22 pairs shows accepted (p-value more than 0.05) Whitney-Mann null test hypothesis. This 84.6% acceptance signify statistically significant relationship between geological formation and soil types to TGR dose rates.

4 Conclusion

From a total of 241 initial measurements taken, 2 outliers were excluded, leaving 239 terrestrial gamma dose rate readings corrected with $C_f = 0.544$, obtained from linear fitting between measured dose rates and calculated dose rates of 29 soil samples taken. Mean corrected dose rate reading D_m is 47 ± 1 nGy h^{-1} with a range between 5 nGy h^{-1} – 103 nGy h^{-1}. This value is lower than the world average of 59 nGy h^{-1} (UNSCEAR, 2000) and much lower than terrestrial gamma dose rates of peninsular Malaysia (165 nGy h^{-1} for Johor, 172 $\pm$ 17 nGy h^{-1} for Melaka, 222 $\pm$ 191 nGy h^{-1} for Kinta, Perak, 176 $\pm$ 5 nGy h^{-1} for Pahang, and 182 $\pm$ 81 nGy h^{-1} for Selangor) (Ramli, Hussein, 2001; Lee et al, 2009; Gabdo et al, 2013; Sanusi et al, 2014) . A prediction model of geological formation and soil types contribution to the dose rate readings was generated, producing the equation $D_{G,S} = 0.847D_G + 0.637D_S - 22.313$ with a normalized Beta coefficient equation of $D_{G,S} = 0.605D_G + 0.395D_S$, translating to 60.5% contribution from geological background and 39.5% contribution from soil types to dose rates. When applied to geology-soil pairings of the study area, the prediction model produces an 84.6% Whitney-Mann acceptance level when tested against corrected dose rate readings, indicating statistically significant relationship between geological formation and soil types to terrestrial gamma dose rates. Thus, the prediction model can be used to predict terrestrial gamma radiation dose rate for areas with similar geology-soil pairing profile across Sarawak.

References

1. A. T. Ramli, A. W. Hussein, (2001). Geological influence on terrestrial gamma radiatio dose rate in the Malaysian State of Johore. *Applied Radiation and Isotopes 54, 327-333.*
2. Directorate of National Mapping Malaysia, (2010). Sarawak Map, edisi 1-PPNM, Director of National Mapping Malaysia. Jabatan Ukur dan Pemetaan Malaysia, Kuala Lumpur.
3. H. T. Gabdo, A. T. Ramli, M. S. Sanusi, M. A. Saleh, N. N. Garba (2013). Terrestrial gamma dose rate in Pahang state Malaysia. *Journal of Radioanalytical and Nuclear Chemistry 299, Issue 3, 1793-1798.*
4. International Atomic Energy Agency (IAEA), (1989). Measurement of Radionuclides in Food and the Environment, A Guidebook. Technical Reports Series No. 295. International Atomic Energy Agency, Vienna.
5. Jabatan Pertanian Kuching Sarawak, (2008). Kumpulan Utama Tanah Sarawak, GIS Cawangan Pengurusan Tanah. Jabatan Pertanian Kuching Sarawak, Sarawak.
6. Jabatan Pertanian Sarawak, (2017, July 14). Betong Divisional Agriculture Office. Retrieved July 14, 2017, from http://www.doa.sarawak.gov.my.
7. M. S. M. Sanusi, A. T. Ramli, H. T. Gabdo, N. N. Garba, A. Heryanshah, H. Wagiran, M. N. Said, (2014). Isodose mapping of terrestrial gamma radiation dose rate of Selangor state, Kuala Lumpur, and Putrajaya, Malaysia. *Journal of Environmental Radioactivity 135, 67-74.*
8. Majlis Daerah Serian, (2017, July 14). Sejarah kami. Retrieved July 14, 2017, from http://www.seriandc.sarawak.gov.my.
9. Majlis Daerah Sri Aman, (2017, July 14). Pengenalan. Retrieved July 14, 2017, from http://www.sriamandc.sarawak.gov.my.
10. Minerals and Geoscience Department Malaysia, (2013). Geological map of Sarawak, third edition, Director General of Minerals and Geoscience Department Malaysia. Minerals & Geoscience Department Malaysia, Kuala Lumpur.
11. Pentadbiran Bahagian Kuching, (2017, July 14). Sejarah Kuching. Retrieved July 14, 2017, from http://www.kuching.sarawak.gov.my.

 DOI: 10.1051/epjconf/201715600006

12. Pentadbiran Bahagian Samarahan, (2017, July 14). Profil Bahagian Samarahan. Retrieved July 14, 2017, from http://www.samarahan.sarawak.gov.my.
13. Pentadbiran Bahagian Sri Aman, (2017, July 14). Profil Daerah Lubok Antu. Retrieved July 14, 2017, from http://www.sriaman.sarawak.gov.my.
14. S. K. Lee, H. Wagiran, A. T. Ramli, N. H. Apriontoro, A. K. Wood, (2009). Radiological monitoring: terrestrial natural radionuclides in Kinta District, Perak, Malaysia. *Journal of Environmental Radioactivity 100, 368-374.*
15. United Nations Scientific Committee on the Effects of Atomic Radiation (UNSCEAR), (2000). Sources, effects, and risks of ionization radiation. Report to the General Assembly, with Scientific Annexes B: Exposures from Natural Radiation Sources. United Nations, New York.
16. Vlado Valkovic, (2000). Radioactivity in environment, Elsevier Science B.V., Amsterdam.

 DOI: 10.1051/epjconf/201715600007

Thermoluminescence (TL) dosimeter of dysprosium doped strontium borate glass for different glass modifiers (Na, Li, Ca) subjected from 1 to 9 Gy doses

S. A. Hamzah[1], *M.A Saeed*[1], *H. Wagiran*[1], and *I. H. Hashim*[1]

[1]Department of Physics, Faculty of Science, Universiti Teknologi Malaysia, 81310 Skudai, Johor Bahru, Malaysia

Abstract. This article reports TL response for different glass modifier and doping concentration. Alkali oxides (Na_2O and Li_2O) and alkali earth oxide (CaO) will be used as a glass modifier for strontium borate based glass. The samples were prepared by melt quenching technique. Dy_2O_3 concentrations ranging from 0.00 to 0.70 mol% and exposure doses of 1 to 9 Gy will be varied. All glass samples exhibit the prominent peak temperature positioned at 186 °C to 232 °C. From all the samples, one of the samples shows an excellent linearity dose response, higher TL and show good reproducibility after 5 cycles exposure which is sodium strontium borate doped with 0.1 mol% Dy_2O_3 (optimum concentration).

1 Introduction

Thermoluminescence dosimeter (TLD) has been widely applied in clinical, personal and environmental of ionizing radiation areas. Borate glass for better dosimeter materials were chosen due to is near tissue equivalent absorption coefficient, high thermal stability, low melting point and good solubility of rare earth ions [1]. The hygroscopic behavior of pure borate glass represents a serious drawback since humidity has often gives contrary effect on the performance of TL materials. Therefore, many investigations have been conducted to improve the performance of borate based dosimeters and explore the new TL materials. Strontium tetraborate is one of the potential candidates since it is not hygroscopic as normally exhibited by other borate compounds [2]. It is reported that the sensitivity of polycrystalline dysprosium doped strontium tetraborate is five times higher compared to LiF: Mg, Ti [3].

Up to now, there are no research study TLD properties on strontium borate glass system. So, in this study strontium borate glass with different modifiers alkali oxides (Na_2O and Li_2O) and alkali earth oxide (CaO) in glassy form and have investigated for the first time its thermoluminescence dosimeter (TLD) properties. Alkali oxide and alkali earth oxides are very good materials for modifying borate glasses because of their possible applications in radiation absorbent. [4,5]. The addition of dysprosium (Dy) as an activator was chosen in this research to enhance the high TL efficiency [6]. The aims of this study is to investigate the TL properties of dysprosium doped with different modifiers (Na_2O, Li_2O and CaO) of strontium borate glasses exposed to ^{60}Co Gamma irradiation. The best modifier of strontium borate glass with optimum concentration of Dy doping will gives the best TL properties such as glow curve, TL dose response, reproducibility, sensitivity to radiation and fading characteristics of the response. A small amount of dysprosium oxide (Dy_2O_3) was added to confirm the role as activator on TL dosimetry response.

2 Materials and Methods

2.1 Sample Preparation

In this study, the glass sample of Dy_2O_3 doped with different modifiers (Na_2O, Li_2O and CaO) of strontium borate at various Dy_2O_3 concentrations ranging from 0 to 0.7 mol% was synthesized. Table 1 shows the nominal compositions of the prepared samples. Melt quenching technique was used to prepare the glasses of compositions (70-x) B_2O_3- 20SrO -10 Na_2O - $x$$Dy_2O_3$, 10$Li_2O$ - 20 SrO - (70-x) B_2O_3 - (x)Dy_2O_3 and 10 CaO - 20 SrO - (70- x) B_2O_3- (x) Dy_2O_3, where, (x = 0.0, 0.1, 0.3, 0.5, 0.7) mol%. The commercially available raw materials in powder form were used. The high pure analar grade chemicals $SrCO_3$ (Aldrich 99.99%), B_2O_3 (Acros Organic 99%), Na_2CO_3 (Merck 99%), Li_2CO_3 (Acros Organic 99%),

$CaCO_3$ (QReC 99.9%) and Dy_2O_3 (Aldrich 99.99%). The required proportions of Na_2O, Li_2O, CaO, SrO, B_2O_3 and Dy_2O_3 powder was weighted using sensitive analytical balance. An appropriate mixture of 20 grams compound was placed in a bottle container. Then the mixture was mixed mechanically by milling machine in a homogenizer for about 1 hour to ensure the homogeneity of the mixture. The mixture was then poured into an alumina crucible and melted in high temperature furnace at 1100 °C for 20 minutes using the melt-quenching technique. The purpose of a high temperature anneal is required to clear the dosimetric traps of the residual signal, which may cause unwanted background during subsequent use of the dosimeters.

After the samples were totally melted, the glass samples were annealed for 4h at 400 °C below their calorimetric glass transition temperature to remove any external stresses and then cooled down to room temperature. Three batch series of samples with different modifiers and concentration of dysprosium dopant were prepared. The Dy_2O_3 as activator enhances the TL performance of the glass samples. After optimum concentration analysis of a single broad peak of TL glow curve is observed and it is located between 180 °C and 250 °C. It is in the same argument with previous research that an ideal glow curve should have a single sharp peak located between 180 °C and 250 °C [7]. The highest TL response with low standard error is observed and recorded for the Dy_2O_3 doped different modifiers of strontium borate glasses. After the analysis, it is found the highest TL response for different concentration Dy dopant were 0.1 mol% Dy doped sodium strontium borate glass, 0.1 mol% Dy lithium strontium borate glass and 0.3 mol% Dy doped calcium strontium borate glass. All of the highest TL response samples were labeled as **N1**, **L1** and **C2** respectively.

2.2 Gamma ray Irradiation

The glass samples were cut into small pieces with approximately same thickness and mass. They were weighted using an electronic analytical balance with an accuracy of 4 decimal places. Then each glass samples irradiated subjected to 50 Gy absorbed dose of ^{60}Co gamma ray at Universiti Kebangsaan Malaysia, UKM, Bangi. After 24 hours of irradiation, the samples were readout using TL reader Harshaw Model 4500 located at Nuclear Laboratory, Physics Department, Faculty of Science, Universiti Teknologi Malaysia. The linear time temperature profile (TTP) of each batch sample were different because different modifiers' presence in strontium borate glass. After few trials of TL readouts using various heating rates, N1 is 3 $^oCs^{-1}$, L1 is 5 $^oCs^{-1}$, and C2 is 7 $^oCs^{-1}$ that gives the best resolution of the glow curve. The TL readings were then normalized to their mass.

3 Result and Discussion

3.1 X- ray Diffraction Analysis

X-ray diffraction analysis (XRD) was used to investigate the amorphous nature of the prepared glass samples. The powder samples of prepared glass were analyzed using Siemens Diffraction D5000, equipped with diffraction software analysis with Cu radiation (= 1.54045 Å) in range $5^o < 2\theta < 80^o$. This equipment is located in Material Science Laboratory, Faculty of Mechanical, Universiti Teknologi Malaysia (UTM). The XRD diffractograms were obtained by one second for each counting time powered at 40 kV and 30 mA at room temperature (300K) by step scanning 0.05^0. **Fig. 1** shows the XRD patterns of the samples. No sharp Bragg peaks are exhibited in any spectrum. The result proves that the samples were purely amorphous and did not exhibit any crystalline phase.

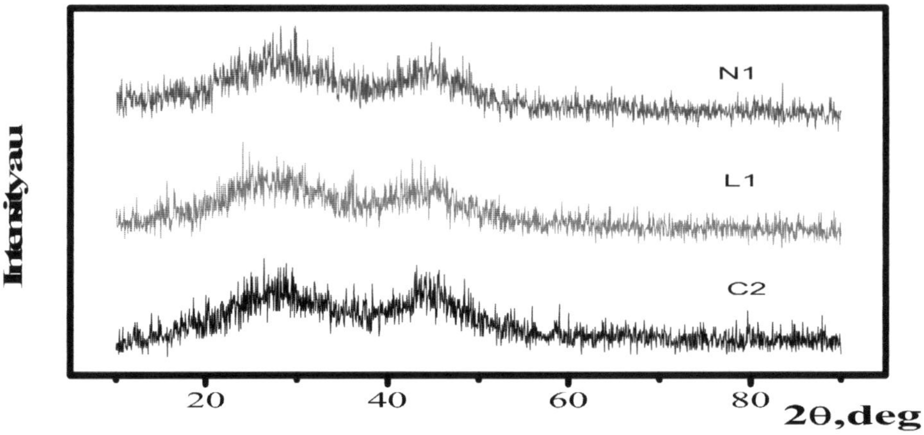

Fig 1:XRD patterns of optimum Dy-doped concentration on different modifiers of strontium borate glass.

3.2 Differential Thermal Analysis (DTA)

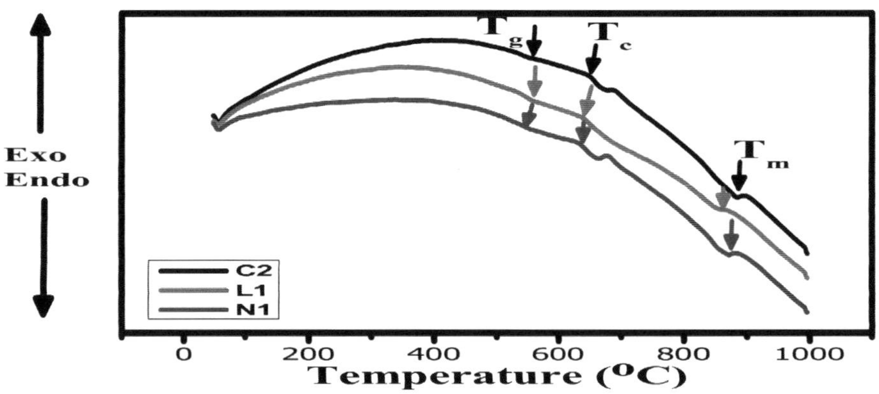

Fig 2: DTA traces of Dy-doped on different modifiers of strontium borate glasses.

Glass forming ability and stability of the prepared glass samples were measured using differential thermal analysis (DTA) curves Pyris STA 8000 analyzer. The endothermic peak corresponding to the glass temperature, T_g, melting temperature, T_m, exothermic peak of the crystallization temperature, T_c and glass transition temperature, T_{rg}, of the glass samples were traced using DTA analysis curves (**Fig. 2**).

Table 1: Thermal stability glass forming ability and Hruby number of the optimum dy-doped different modifier of strontium borate glasses

Sample No.	$T_g \pm 0.01(^oC)$	$T_c \pm 0.01\ (^oC)$	$T_m \pm 0.01(^oC)$	$\Delta T \pm 0.01(^oC)$	$T_{rg} \pm 0.01(^oC)$	$H_g \pm 0.01(^oC)$
N1	542.69	653.96	864.80	111.27	0.63	0.53
L1	549.99	637.88	854.52	87.89	0.64	0.41
C2	551.18	671.30	881.11	120.12	0.63	0.58

Glass forming ability is estimated using Kauzmann relation:

$$T_{rg} = T_g / T_m \quad (1)$$

In this study, the values of T_{rg} appear in the range of 0.63 to 0.64 and Hurby relation, H_g varies from 0.53 to 0.58. The calculated glass forming ability and thermal stability are summarized in **Table 1**. Glass forming ability is said to be good if it is in the range $1/2 \leq T_{rg} \leq 2/3$. Kauzmann criterion is well-obeyed for the present glass composition. Glass stability is said to be very poor for $H_g \leq 1$ and superior for $H_g \geq 0.5$.

3.3 TL dose response

The TL response of sample (L1,N1 and C2) were exposed from 1 Gy to 9 Gy doses range of gamma irradiation. Every data point represent an average of five individual glass readings. The TL response is slightly energy dependent, and the value of TL response tends to spread wider at higher energies from 1 Gy up to 9 Gy doses. **Fig. 3** show that sample C2 gives a linear and the highest TL dose response at high energies compared to sample N1 and L1. As the dose range increases from 1 Gy to 9 Gy, the TL intensity of glass sample N1 also increases linearly.

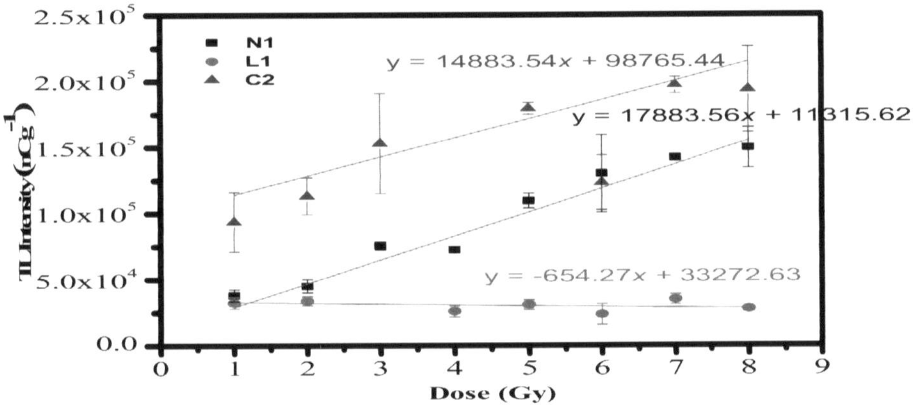

Fig. 3. TL dose response for low dose range from 1 Gy to 9 Gy doses range.

3.4 Sensitivity

Table 2: Sensitivity and the relative sensitivity of low dose energies of N1, L1, and C2 glass samples from 1 to 9 Gy.

TLD Material	Range of Dose	Sensitivity ($nC\ g^{-1}\ Gy^{-1}$)	Regression Coefficient, R^2	Relative Sensitivity with Respect to TLD-100
N1		0.018×10^6	0.92	0.016
L1	1 to 9 Gy of ^{60}Co gamma ray	0.0065×10^6	0.19	0.00058
C2		0.015×10^6	0.85	0.013
TLD-100		1.12×10^6	0.93	1

TL sensitivity is defined as TL intensity signal over certain temperature region per unit absorbed dose per unit mass ($nC\ g^{-1}\ Gy^{-1}$). The TL sensitivity signal can directly determine from the slope of the graph of TL intensity response as a function of dose. **Table 2** summarized sensitivity and the relative sensitivity of low dose energies of N1, L1, and C2 glass samples from 1 to 9 Gy doses. From our observation, sample N1 has the highest relative sensitivity. It shows that the sensitivity sample. N1 is slightly near to TLD-100. TL materials with high sensitivity allow for their utilization in the assessment of low doses [8,9].

3.5 Reproducibility

Reproducibility is one of the essential characteristics of TL properties to determine the reusability of the glass samples. Thus, TL intensity signal of a particular dose should be approximately the same after several measurements. A TL material is expected to not suffer any physiochemical change due to repeated irradiation, readout and annealing process [10]. **Fig 5** parade different samples (N1, L1, C2) at a dose 5 Gy of gamma irradiation after five times sequential cycles. There is no difference between the results of the sequential measurement of low dose at 5Gy. The result confirms that these glass materials are reusable in radiation dosimeter.

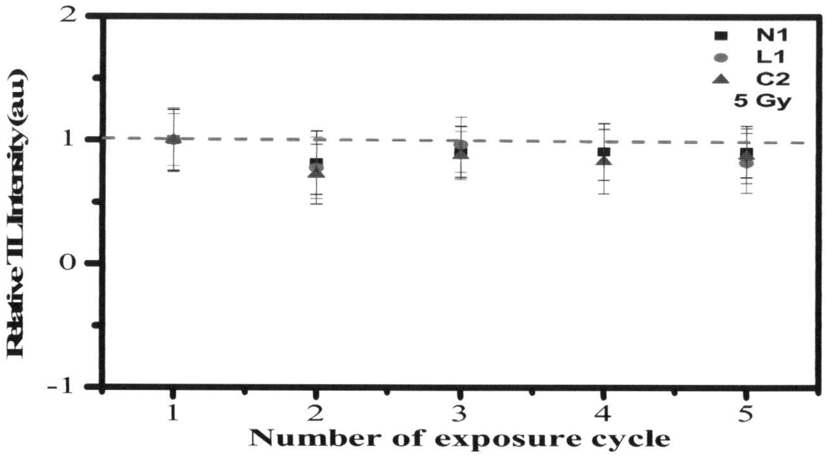

Fig 5. Reproducibility of different glass samples irradiated at 5 Gy doses of gamma irradiation.

3.6 Thermal Fading

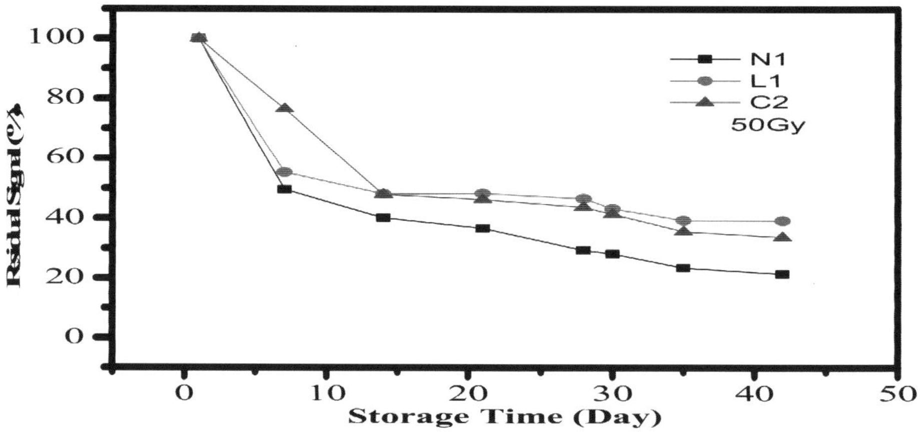

Fig 6. Thermal fading pattern after 42 day of exposure at 5 Gy for L1, N1 and C2.

Thermal fading is a reduction of the TL intensity signal due to the effect of ambient temperature after a period [11]. The fading behavior of samples (N1, L1, C2) up to 42 day is illustrated in **Fig. 6**. After irradiation, the samples were stored in a dark room at room temperature to minimize the possible escape of trapped electrons due to the simulation of temperature and time. It can be seen the TL loss signals after 42 day. It shows that the TL signal rapidly drops in the first 7 days and decrease slowly until day 42 are observed in every sample. Therefore, the samples need to be readout immediate after irradiation.

4. Conclusion

The basic TL dosimeter properties of strontium borate glass system doped with various concentration of Dysprosium (Dy) has been investigated subjected to 1 Gy to 9 Gy doses range of ^{60}Co gamma irradiation. The optimum concentration of Dy has been identified for $NaSr(BO_3)_3$, $LiSr((BO_3)_3$ and $CaSr((BO_3)_3$ are 0.1 mol% (N1), 0.1 mol% (L1) and 0.3 mol% (C2). The XRD analysis confirms the amorphous state of the glass samples and DTA analysis shows for each glass samples in a stable state. The variation of TL dose response, sensitivity, reproducibility and thermal fading has been identified for all batch series of glass samples. Comparison on the three modifiers shows that sodium strontium borate glass doped with 0.1 mol% of dysprosium (N1) oxide gives the best TL dosimeter response. It proves that the addition of different modifier influences the TL dosimeter response.

Acknowledgment

The authors would like to thank the Ministry of Higher Education (MOHE) of Malaysia and Universiti Teknologi Malaysia (UTM) for supporting the research and providing research grant (Q.J130000.2526.12H75). The authors also would like to take the opportunity to thank Universiti Kebangsaan Malaysia (UKM), Bangi for providing irradiation facilities.

References

[1] L.H. Jiang, Y.L. Zhang, C.Y. Li, J.Q. Hao, Q. Su,. Appl. Radiat. Isot. **68**, 196-200 (2010).
[2] M. Santiago, C. Grasseli, E. Caselli, M. Lester, A. Lavat, F. Spano. Phys. Status Solidi A. **185 (2),** 285-289 (2001).
[3] A. Lavat, C. Graseli, M.Santiago, J. Pomarico, E. Caselli, Cryst. Res. Technol. **39 (10)**, 840-848 (2004).
[4] C. Gautam, A. K. Yadav, A. K. Singh, ISRN Ceramics. 2012.
[5] H., Aboud, H. Wagiran, I. Hossain, R. Husin, Int. J. Phys. Sci. **7(6)**, 922-926 (2012).
[6] M. Prokic, E. Yukihara, Dosimetric characteristics of high sensitive Mg2SiO4: Tb solid TL detector. Radiat. Meas. **43(2)**, 463-446 (2008).
[7] T.Y. Lim, H. Wagiran, R. Husin, S. Hashim, Appl. Radiat. Isot **102**, 10-14 (2015).
[8] Y. Alajerami, PhD, Universiti Teknologi Malaysia (2014).
[9] J. Marcazzo, M. Santiago, C. D'Angelo, C. Furetta, E. Caselli,. Nucl Instrum Methods Phys Res B. **268(2)**, 183-186 (2010).
[10] J. Manam, S.K. Sharma, Nucl. Instrum. Methods Pyhs. Res. B. **217(2),** 314-320 (2004).
[11] S.W.S.,McKeever, M. Moscovitch, P.D. Townsend Nucl. Technology Pub. Ashford, Kent, England (1995).
[12] C. Furetta,. *Handbook of Thermoluminescence.* Singapore: World Scientific, (2003).
[13] C. Furetta, M. Prokic, R. Salamon, G. Kitis, Appl. Radiat. Isot. **52**, 243-250 (2000).
[14] C. Furetta, M. Prokic, R. Salamon, G. Kitis, Nucl. Instr. Meth. Phys. Res. A. **456 (3),** 411-417 (2001).
[15] L.H. Jiang, Y.L. Zhang, C.Y. Li, J.Q. Hao, Q. Su, J. Mater. Sci. **61**, 5107-5109 (2007).
[16] L.H. Jiang, Y.L. Zhang, C.Y. Li, J.Q. Hao, Q. Su,.J. Alloy Mater. **482**, 313-316 (2009).
[17] S. Samat, C. J. Evans, Satistics and Nuclear Counting: Theory, Problems and Solutions (1992).
[18] K. O. Hayder, R. Hussin, H. Wagiran, M. A. Saeed. J. Non-Cryst. Solids, **427**, 83-90 (2015).

EPJ Web of Conferences **156**, 00008 (2017)
RCNP 2016
DOI: 10.1051/epjconf/201715600008

Environmental radiology assessment in Lahad Datu, Sabah

Yusof Siti Fharhana.[1,*], *Wan Hassan* Wan Muhamad Saridan[1], *Ramli* Ahmad Termizi[1], *Sahini* Mohd Hilmi[1], *Mohd Sanusi* Mohammad Syazwan[1], *Basri* Nor Afifah[1], and *Abu Hanifah* Nor Zati Hani[1]

[1]Department of Physics, Faculty of Science, Universiti Teknologi Malaysia, 81310 UTM Johor Bahru, Johor, Malaysia

Abstract. Monitoring terrestial gamma radiation is crucial to prepare a baseline data for environmental radiological protection. Radiological research was carried out in Lahad Datu, Sabah to obtain the radioactivity status and terrestrial gamma radiation level in the area. We measure the terrestrial gamma radiation dose rates and analyse the radioactivity concentration of primordial radionuclides for radiological risk assessment. We identified that the annual estimation of dose effective for public is below the public dose limit, 1 mSv per year. Public and environment safety and health are remain secure. The obtained data and results can be used as reference for environmental radiology protection.

1 Introduction

1.1 Research Background

Due to its ionising power, high terrestrial gamma radiation dose may cause a detrimental effect to the public. The dilemma faced by public, governments and international radiation protection community on radiation health effects and risk of low doses has lead to the studies in the high background natural radiation area. Since the low level exposures may bring a stochastic effects to the public, the study of background radiation is extremely vital [1].

Public exposed to the radiation that occur from natural sources. Human beings were estimated to receive about 87% natural radiation. Therefore, it is important to assess environmental radiology for health and safety purpose [2].

Since the damage of Fukushima Daiichi nuclear plant tragedy, it is reported that 80% of radionuclides has entered the Pacific Ocean [3]. Therefore, this study should be conducted as in Peninsular Malaysia because the Sabah is surrounded by ocean, such as the South China Sea, Sulu Sea, the Celebes Sea and the Pacific Ocean.

*Corresponding author: sfharhana@gmail.com

1.2 Problem Statement And Research Objectives

This research is conducted to measure the terrestrial gamma radiation in Lahad Datu district. This is also to prepare baseline data of terrestrial gamma radiation. Data collection in Peninsular Malaysia has been conducted but not in Lahad Datu, Sabah [18]. The accident at the Fukushima Daiichi Nuclear Power Plant in Japan in March 2011 resulted in the release of several hundred pBq of activity into the environment [5]. Therefore this research will identify the concentration of primordial radioanuclide for radiological health and safety assessment. Malaysia chose nuclear energy to guarantee supply beyond 2030 [6]. Baseline data is one of important aspect in assessing nuclear power plant site [7]. Hence, this research will build an isodose map of ionizing radiation in Lahad Datu for further reference.

2 Methodology

2.1 The study area

Lahad Datu district is located in southeastern Sabah, coordinated between latitudes 4° 45' N and 5° 30' N, and longitudes 117° 30' E and 119° 15' E. It covers an area of 6,501 km^2 with 195,865 population in 2010. This district is a great socio-economy contributor to Sabah State as it has the largest palm tree cultivation [8] where the world-class Palm Oil Industrial Cluster has been built here [9]. The world-class tourism spot, Danum Valley Conservation Area and Tabin Wildlife Reserve are also located in this district [9].

2.2 Measurements of terrestrial gamma radiation dose rate

The sampling points were determined based on the types of soil and geology in research area as the terrestrial gamma radiation is vary depends on the radionuclides concentration in soil [10]. The terrestrial gamma radiation dose rate was measured by using a survey meter (Ludlum19 micro R meter). Measurements were conducted at approximately 1 m above the surface of the land [11].

2.3 Analysis of radioactivity concentration of primordial radionuclides

The collected soil samples is prepared in Marinelli beaker for the analysis of natural concentration of radionuclide by using HPGe gamma spectrometer detector in Nuclear Laboratory, Faculty of Science, UTM. The analysis of gamma radiation spectrum is carried out by using gamma radiation counter system (GC2018-7500SL model) paired with Genie 2000 VI.3 software. The concentration of ^{238}U is calculated based on the energy peak of gamma ray of daughter radionuclides ^{214}Pb and ^{214}Bi in the ^{238}U decay series. The concentration of thorium is calculated based on the energy peak of gamma rays of daughter radionuclides ^{228}Ac and ^{208}Tl in ^{232}Th decay series. ^{40}K concentration is calculated based on the gamma ray emission peak of 1460.8 keV energy [12].

2.4 Radiological risk assessment

After analysing the concentration of natural radioactivity, radiological risk assessment is carried out by calculating the Annual Estimation of Dose Effective, AEDE, the estimated probability of occurrence of cancer risk, G, the Radium Equivalent Activity, Ra_{eq}, and the Hazards Index, H_{ex}.

2.5 Isodose mapping

Isodose map of gamma irradiation in Lahad Datu is built based on terrestrial gamma radiation survey by using spatial mapping analysis software ArcGIS 9.3.

3 Results and discussion

3.1 Terrestrial gamma dose rates

A total of 86 sampling points were determined and the reading of gamma radiation dose rates was measured by using survey meter. Out of 86 sampling points, 10 soil samples labeled as S1 to S10 were taken for the analysis of the concentration of radioactivity. The soil samples were taken as representatives of different doses of dose levels; low and high doses relative to the environmental level. The highest gamma dose rate is 79.4 nGy h^{-1}. The lowest terrestrial gamma dose rate is 6.1 nGy h^{-1}. The mean terrestrial gamma dose rate is 44.7 nGy h^{-1} much lower than the mean terrestrial gamma dose rate proposed by UNSCEAR (2000), 57 nGy h^{-1}.

3.2 Radioactivity concentration of primordial radionuclides

Table 1 illustrates the radioactivity concentration of primordial radionuclides, C in soil samples. The mean value of C_{U-238}, C_{Th-232} and C_{K-40} is 42.6 Bq kg^{-1}, 77.7 Bq kg^{-1} and 810.9 Bq kg^{-1} respectively. It was assumed that the high value of potassium concentration is due to the contribution of potassium content in fertilizer used in palm tree cultivation.

Table 1: Radioactivity concentration of ^{238}U, ^{232}Th and ^{40}K.

Sample	C_{U-238} (Bq kg^{-1})	C_{Th-232} (Bq kg^{-1})	C_{K-40} (Bq kg^{-1})
S1	46.2	87.1	961.0
S2	48.8	78.4	817.0
S3	28.5	46.7	473.0
S4	47.9	80.0	764.0
S5	58.1	110.7	1225.7
S6	43.2	67.2	747.0
S7	57.0	107.9	900.0
S8	55.3	105.4	1200.0
S9	18.7	34.0	426.0
S10	22.7	59.3	595.2
Mean concentration	42.6	77.7	810.9

3.3 Radiological risk assessment

Using equation (1), the annual effective dose equivalent, AEDE to the public was calculated (UNSCEAR, 2000).

$$\text{AEDE} = \text{Dose rate nGy j}^{-1} \text{ x 24 hrs x 365 days x 1 OF x 0.7 Sv Gy}^{-1} \text{ x } 10^{-6} \quad (1)$$

Annual Estimation of Dose Effective, AEDE is 0.44 mSv per year lower than public dose limit, 1 mSv per year.

The probability of the risk of cancer , G_r is estimated by using equation (2). The risk factor, R is 0.05 Sv^{-1} and H is the AEDE value [13].

$$G_r = R.H \quad (2)$$

The estimated probability of the risk of cancer, G_r is 0.02 probability per year per person. The value of the probability of cancer risk is too small to be considered as radiological hazard to the public in the study area.

Radium equivalent activity , Ra_{eq} is calculated by using equation (3) [14].

$$Ra_{eq} = C_{U\text{-}238} + 1.43(C_{Th\text{-}232}) + 0.0077(C_{K\text{-}40}) \quad (3)$$

Radium equivalent activity, Ra_{eq} is 160 Bq kg^{-1} which is less than the recommended limit of concentration for TENORM material which is 370 Bq kg^{-1}[15].

Equation (4) is used to calculate the hazard index, H_{ex} [15].

$$H_{ex} = \frac{C_{U-238}}{370} + \frac{C_{Th-232}}{259} + \frac{C_{K-40}}{4810} \leq 1 \quad (4)$$

The value of H_{ex} must be lower than 1 which equivalent to the maximum recommended limit of radium activity, 370 Bq kg^{-1} [16]. In this study, the H_{ex} value is 0.58 which is lower than 1.

3.4 Isodose mapping

Figure 1 exhibits an isodose map based on the survey of gamma radiation in the field. The map contour generated in the isodose map based on the surveys are still rough where there is a large gap between the contour for the less observed area. This is because the accuracy of the interplotation data is proportional to the distance between dose rate survey points [17]. By following the dose rate range of isodose maps of Peninsular Malaysia [18], Figure 1 shows a very low dose rate in research area which indicates a safe environment for public.

DOI: 10.1051/epjconf/201715600008

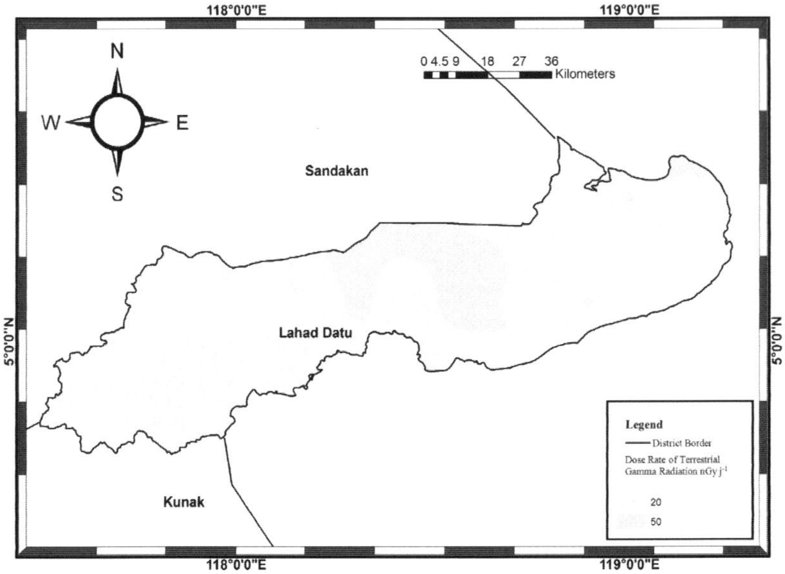

Figure 1: Isodose map based on gamma radiation survey in study area.

4 Conclusion

The level of gamma radiation in the environment is too small to cause an acute effect and macro stochastic effect. We identified that the annual estimation of dose effective for public is below the public dose limit, 1 mSv per year. Therefore, public and environment safety and health are remain secure. The obtained data and results from this study can be used as reference for environmental radiology protection.

The authors wish to acknowledge the Ministry of Science, Technology and Innovation (MOSTI) and Universiti Teknologi Malaysia for supporting this research by providing R.J130000.7926.4S109 grant.

[1] M. Sohrabi, World high background natural radiation areas: Need to protect public from radiation exposure, **50**, 166-171, (2012).
[2] N. Ahmad, M. S. Jaafar, M. Bakhash, M. Rahim, An overview on measurements of natural radioactivity in Malaysia, **8**, 136-141, (2015).
[3] O. Evrard, J. P. Lepage, Y. Onda, O. Cerdan, S. Ayrault, Radiocesium transfer from hillslopes to the Pacific Ocean after the Fukushima Nuclear Power Plant accident: A review, **148**, 92-110, (2015).
[4] Department of Agriculture Sabah, Hectareage of industrial crops by districts and types of crops, Sabah, (2014), retrieved on 22 April 2016 from http://ww2.sabah.gov.my/
[5] Barletta, W. A., Stoop J, O'Hara C., Bailiff, I. H., Hermanspahn, N., Hugtenburg, R., Alcock, R., Covaci, A., Barcelo, D., Gan, J. and Karlsson, (2016). 5 years after Fukushima — insights from current research. A virtual special issue featuring free access to impactful research from a multidisciplinary perspective.
[6] Basri, N. A. and Ramli, A. T., (2012). Selection of possible candidate area for nuclear power plant in Johor, Malaysia. Nuclear and Related Technologies, 9 (1). pp. 56-63. ISSN 1823-0180

[7] Saleh, M. A., Ramli, A. T., Alajerami Y., Mhareb M. H. A., Aliyu A. S., Gabdo H. T., Garba N. N. (2014). Assessment of radiological health implicat from ambient environment in the Muar district , Johor , Malaysia. Radiation Physics and Chemistry. 103, 243-252.
[6] Lahad Datu District Office, Latar Belakang PDLD, retrieved on 11 Jun 2015, from http://ww2.sabah.gov.my/
[7] Z. Wang, Natural radiation environment in China, **1225**, 39-46, (2002).
[8] UNSCEAR. United Nations Scientific Committee on the Effect Atomic Radiation (2000). *Sources and Effects of Ionizing Radiation*. UNSCEAR Report on The General Assembly. New York : United Nations.
[9] G. Wallova, N. Kendler, G. Wallner, Monitoring of radionuclides in soil and bone samples from Austria, **107**, 44-50, (2012).
[10] International Commission on Radiological Protection ICRP 60, (1990).
Recommendations of the international commission on radiological protection. In: ICRP Publication 60, Annals of the ICRP. Pergamon Press,
Oxford.
[11] Y. Orgun, N. Altınsoy, S.Y. Sahin,Y.Gungor, A.H. Gultekin, G. Karahan, Z. Karacık, Natural and anthropogenic radionuclides in rocks and beach sands from Ezine region (C - anakkale), Western Anatolia , Turkey, **65**, 739-747, (2007).
[12] E. O. Agbalagba, G. O. Avwiri, Y. E. Chad-umoreh, g -Spectroscopy measurement of natural radioactivity and assessment of radiation hazard indices in soil samples from oil fi elds environment of Delta State , Nigeria, **109**, 64-70, (2012).
[13] A. Sh.Aydarous, S. Zeghib, M. Al-Dughmah, Measurements of natural radioactivity and the resulting radiation doses from commercial granites, **142**, 363-368, (2010)
[14] A. T. Ramli, M. S. M. Sanusi, N. A. Basri, Final Report; Isodose Mapping of Terrestrial Gamma Radiation in Peninsular Malaysia, UTM Skudai, (2013)

Radioactive dispersion analysis for hypothetical nuclear power plant (NPP) candidate site in Perak state, Malaysia

Shazmeen Daniar Shamsuddin*[1], Nor Afifah Basri[2], Nurlyana Omar[3], Meng-Hock Koh[4], Ahmad Termizi Ramli[5], and Wan Muhamad Saridan Wan Hassan[6].

[1,2,3,4,5,6] Department of Physics, Faculty of Science, Universiti Teknologi Malaysia, 81310, Skudai, Johor, Malaysia.

*shazmeendaniar@gmail.com

Abstract. Malaysia is planning to build a nuclear power plant (NPP) by 2030 to diversify the national electricity supply and resources. Selection of an NPP site must consider various factors, especially nuclear safety consideration to fulfil the nuclear safety objectives. Environmental Risk Assessment Analysis is a part of safety requirements by the International Atomic Energy Agency (IAEA) prior to the NPP commissioning process. Risk Assessments Analysis (RIA) is compulsory for the NPP site evaluation. One of RIA methods are Radioactive Dispersion Analysis using probabilistic risk analysis software. It is also important to perform studies to estimate the impact to the neighbouring population in the case of a nuclear accident at the power plant. In the present work, aimed to study the impact of a hypothetical nuclear accident by simulating the dispersion pattern of radionuclides originated from a candidate site at Manjung, Perak. The work has been performed using the HotSpot Health Physics codes. Two types of radionuclides have been considered namely ^{137}Cs and ^{131}I. In calculations, the initial concentration of radioactive materials of Fukushima Daiichi accident data are used which are 2.06 x 10^{16} Bq and 1.68 x 10^{17} Bq respectively for the two radionuclides. The result shows that the dispersion distance obtained from both software are not the same. It shows that ^{137}Cs and ^{131}I can be dispersed as far as 16 km and 80 km away from the site during radiological accident respectively, reaching major towns in Perak. Using HOTSPOT, the estimated total effective dose equivalent (TEDE) for ^{137}Cs and ^{131}I at major towns in Perak such as Lumut and Sitiawan are 1.2 mSv and 9.9 mSv. As for Taiping, Ipoh, Kampar, and Teluk Intan the estimated TEDE is around 0.2 mSv and 1.6 mSv respectively. In conclusion, the dispersion can reach as far as 80 km from the site. However, estimated annual effective dose is not more than 1 mSv limit, which is considered acceptable in the point of view of radiological health risk for human and the environment.

Keywords; nuclear power plant, radioactive, dispersion, accident

1. Introduction

Nuclear power is one of energy source in electricity sector which represent 13% to 15% of global electricity production. As of 2014, total nuclear electricity production is 2,410 TWh around the world [1].

Due to ever increasing demand of electricity, many non-nuclear countries have shown interest to include nuclear in their energy mix, including Malaysia. However, the driving forces are different in every emerging nuclear country [2]. In Malaysia's case, the ultimate driving force is to overcome over-dependence on fossil fuels and to serve as an alternative source to supply base load demand. Current scenario in Malaysia shows that 130 billion kWh gross was produced in 2011, 58 billion kWh (45%) of this from gas, 53 billion kWh (41%) from coal, 7.6 billion kWh (6%) from hydro and 10 billion kWh (8%) from oil. In 2011, the Energy Commission showed 28.4 GWe capacity, which is 51.6% of this supplied by gas and 27% by coal [3][4].

The decision to consider nuclear to be included in national energy mix was made in 2009. In June 2009, the government formally decided to consider nuclear power after a comprehensive energy policy study has been done. As to plan and coordinate the nuclear power development program, a Nuclear Power Development Steering Committee was set up, via three groups [3]. The pre-project activities were conducted discreetly by various government agencies, until the plan to build two units of 1000 MW of nuclear power plant was formally announced in 2010. The government has established Malaysian Nuclear Power Corporation (MNPC) to lead the governmental agencies and industrial organization in the preparation activities. Series of studies are conducted to systematically examine and evaluate the national state-of-preparedness for the implementation of nuclear power program in Malaysia [4]. In May 2010, with a $7 billion budget which the government had to build a nuclear power plant, the Ministry of Energy, Green Technology and Water was given the task to find a suitable site so that the first unit could be operating by 2021. Thus, five possible locations have been identified in Peninsular Malaysia [3]. However, those locations are not revealed to the public.

Strict regulatory requirements must be fulfilled before a construction license is issued for a new NPP [5]. Site selection of nuclear power plant (NPP) is one of the compulsory activity in the feasibility study of nuclear technology development before a construction license was obtained. Selection of an NPP site must consider various factors, with special deliberation regarding the nuclear safety to fulfil the nuclear safety objectives stated in the international licensing standards [6]. The main objective is to protect the civilian and the environment from radiological risks of nuclear activities.

In the process of site selection, risk assessments are essential regulatory standards for the NPP site evaluation [7]. These assessments include the estimation of potential risks from normal routine and accidental releases of radionuclides from the new NPP to the surrounding. One of risk assessment methodologies is Radioactive Dispersion Analysis (RDA), which is used to estimate human health and environmental impacts from radionuclide dispersion and deposition in the region in normal operation or radiological accident scenario.

This paper presents the atmospheric RDA from hypothetical NPP candidate site in Perak State, Malaysia using HotSpot health physics code. The aim is to demonstrate how the software can be applied for radiological risk assessment of routine and accidental releases from the NPP.

2. Model Description

Radioactive dispersion and deposition in the atmosphere is usually analysed using Gaussian plume model. HotSpot use this model to project the dispersion and deposition of radionuclide in the atmosphere.

2.1 Gaussian plume model

The Gaussian dispersion equation [8] is shown in Equation 1 below;

$$C(x,y,z) = \frac{Q}{2u\pi\sigma_y\sigma_z}\exp\left(\frac{-y^2}{2\sigma_y^2}\right)\left[\exp\left(\frac{-(z-H)^2}{2\sigma_z^2}\right)+\exp\left(\frac{-(z+H)^2}{2\sigma_z^2}\right)\right]$$

(Eq.1)

Where,

C = concentration at a given position (μgm^{-3}),
Q = source emission rate (gs^{-1}),
x = downwind (m),
y = crosswind (m),
z = vertical direction (m),
u = average wind speed (ms^{-1}) at the H
H = height of the release (m).

While σ_y *and* σ_z *are the standard* deviations of the concentration distribution and frequently referred as dispersion or diffusion coefficient which indicates the spread of the plume in the y and z directions (m) respectively. Normally, the values of σ_y *and* σ_z *are determined based on the stability of the atmosphere and the values increases as the distance x increases from the source.*

For the past decades, Gaussian dispersion models have become a unique and efficient tool for the management of air quality. Thus, this model has been widely and successfully used for a wide range of studies of air quality in urban and industrial areas.

2.2 HotSpot Health Physics Codes

The HotSpot Health Physics Codes, or HotSpot program, gives a first-order estimation of the radiation impacts connected with the atmospheric release of radioactive materials. The program was made for the emergency response personnel and planner to prepare a quick, field-convenient arrangement of programming apparatuses to asses occurrences related to radioactive material.The product is additionally utilized for safety-analysis examination of facilities taking care of radioactive material.

The atmospheric dispersion models from this software are designed for short-range distance which is less than 10 km, and short-term time which is less than a few hours of predictions in unobstructed terrain and simple meteorological condition. As stated by Homann and Aluzzi in 2014, HotSpot program provide a fast and usually conservative means for estimation of the radiation effects related to the atmospheric release of radioactive materials [9].

3. Study Area and Atmospheric Data

3.1 Study Area

Perak State is in the West of Peninsular Malaysia. Perak has total area of 21,035 km^2 with total population of 2,460,800 in 2010 [10]. The hypothetical NPP candidate site (will be referred to as CS1 throughout this paper) is in Manjung District (Coordinate - 4.324108 N, 100.564970 E). This site has less population than the entire district, hence may be suitable for an NPP site.

Manjung is the second smallest district in Perak State which covers an area of 1,074 km^2. The nearest town are Lumut and Sitiawan, located around 18 km South East of the site. Total population in Manjung is 239,400 in 2010 with population density of 220 per km^2 [10]. Manjung's property is dedicated generally to agriculture, with an agrarian region of 833.75

DOI: 10.1051/epjconf/201715600009

km^2 or around 71.20% of Manjung District. Forest reserves represent another 168.81 km^2 (14.42%), while residential area is around 29.32 km^2 (2.50%), and swampy region is 68.57 km^2.

3.2 Atmospheric Data

Generally, Perak state has warm and sunny days, while its nights are cool the whole year through, with occasional rain in the evenings. Temperature in Perak ranges from 23°C to 33°C, with humidity more than 82.3 percent. The annual rainfall in Perak measures about 3,200 mm.

Atmospheric RDA analysis in this study utilizes meteorological data in year 2013 of Manjung district. The analysis of meteorological data for CS1 shows no significant difference compared to other areas in Peninsular Malaysia. The average wind speed is 3 to 5 m s^{-1} during non-monsoon season and 5 to 7 m s^{-1} during North East Monsoon (November to March). The maximum wind speed during 2013 is in October with the reading of 7.2 m s^{-1}. October is an inter-monsoon month, where the wind direction turned from Southwest direction to the opposite direction.

As shown in Figure 1 below, the most frequent wind direction in CS1 blows towards the East, throughout the year 2103. 13.8% from total wind speed blows from the West direction in the range of 0.01 to 8.23 m s^{-1}.

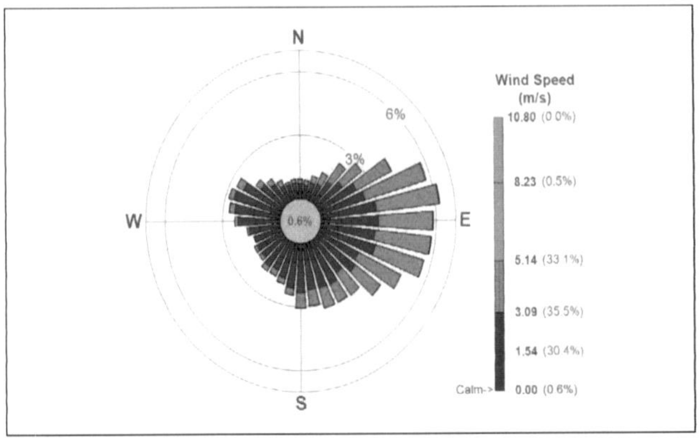

Figure 1. Wind Rose showing the direction of wind blew to and wind speed in 2013

While the wind data determined the direction of radioactive transport over the lands in east-central Malaysia, the rainfall data will determine the deposition of radionuclides to the ground.

4. RDA Methodology

This atmospheric RDA was conducted based on a postulated radiological accident scenario. The accident scenario is set to be equivalent to Fukushima Accident in 2011. Fukushima Accident is the most recent radiological accident due to failure of cooling system after a series of earthquakes, followed by tsunami at the Fukushima Dai-ichi NPP [11][12].

4.1 Dispersion in Postulated Accident

The dispersion simulation of a postulated radiological accident was conducted with reference to the studies on Fukushima Accident 2011 [11][12][13]. The meteorological condition was set to the most frequent meteorological condition in the site, while the material at risk was set equivalent to Fukushima Accident to simulate the same scenario on the site. The parameters set for the simulation are stated in Table 1 below;

Table 1: Parameters for HotSpot Simulation

Parameter	Features	Notes
Dispersion Model	General Plume	-
Radionuclide (Material at Risk)	I-131 and Cs-137	[11] [12] [13]
Deposition Velocity	0.2 cm/sec	[12]
Material at Risk	Cs-137: 2.06 x 10^{16} Bq I-131: 1.68 x 10^{17} Bq	[12]
Sample time	60 minute	[12] [13]
Wind data	2.33 ms^{-1}	-
Source height	20 m	[12] [13]
Meteorological Condition	Stability class B	Low sun and cloudy - average meteorological condition in Malaysia

5. Result on Postulated Accident

From the meteorological data, the information was inserted in the HotSpot software to simulate the dispersion of Cs-137 and I-131 in CS1 by referring to Fukushima Accident scenario (Table 1).

Figures below shows the dispersion shape of Cs-137 and I-131 in the region. Since Hotspot can only simulate one wind direction at a time, 4 directions were selected to observe the dispersion shape towards major towns (Taiping, Ipoh, Kampar and Teluk Intan) in Perak (Figures 2 and 3). Dose conversion data for each radionuclide were listed in Table 2 and 3.

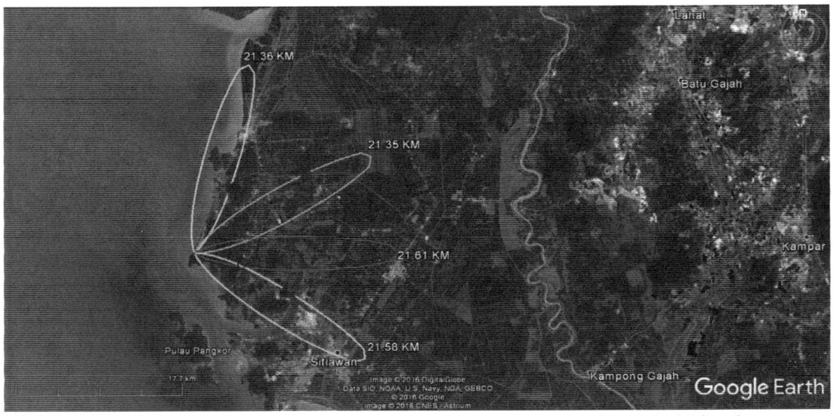

Figure 2: Cs-137 Total Effective Dose Dispersion in the Region in 4 Wind Direction

DOI: 10.1051/epjconf/201715600009

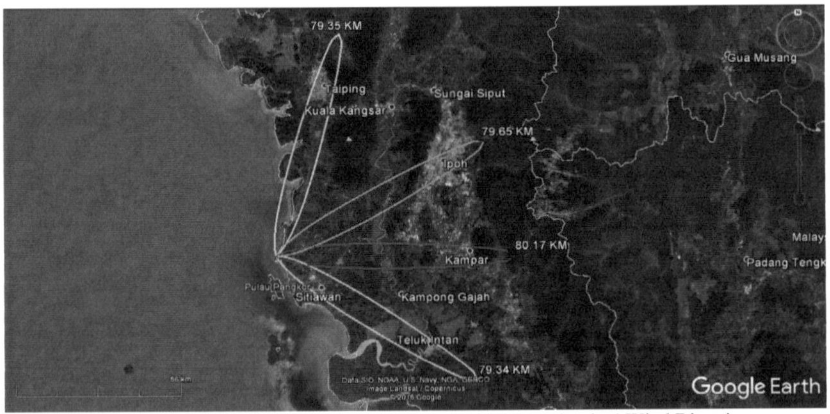

Figure 3: I-131 Total Effective Dose Dispersion in the Region in 4 Wind Direction

Table 2: HotSpot Dose Conversion Data for I-131

DISTANCE km	T E D E (Sv)	RESPIRABLE TIME-INTEGRATED AIR CONCENTRATION (Bq-sec)/m3	GROUND SURFACE DEPOSITION (kBq/m2)	GROUND SHINE DOSE RATE (Sv/hr)	ARRIVAL TIME (hour:min)
0.030	2.5E-03	8.2E+08	3.5E+02	4.8E-07	<00:01
0.100	6.0E+01	2.0E+13	4.0E+07	5.4E-02	<00:01
0.200	4.2E+01	1.4E+13	2.8E+07	3.8E-02	00:01
0.300	2.3E+01	7.7E+12	1.5E+07	2.1E-02	00:02
0.400	1.4E+01	4.6E+12	9.2E+06	1.3E-02	00:02
0.500	9.1E+00	3.1E+12	6.1E+06	8.3E-03	00:03
0.600	6.5E+00	2.2E+12	4.3E+06	5.9E-03	00:04
0.700	4.8E+00	1.6E+12	3.2E+06	4.4E-03	00:04
0.800	3.7E+00	1.3E+12	2.5E+06	3.4E-03	00:05
0.900	3.0E+00	1.0E+12	2.0E+06	2.7E-03	00:06
1.000	2.4E+00	8.1E+11	1.6E+06	2.2E-03	00:06
2.000	6.4E-01	2.1E+11	4.3E+05	5.8E-04	00:13
4.000	1.7E-01	5.8E+10	1.2E+05	1.6E-04	00:27
6.000	8.1E-02	2.7E+10	5.5E+04	7.4E-05	00:40
8.000	4.8E-02	1.6E+10	3.3E+04	4.4E-05	00:54
10.000	3.3E-02	1.1E+10	2.2E+04	3.0E-05	01:08
20.000	9.9E-03	3.3E+09	6.7E+03	9.0E-06	02:16
40.000	3.2E-03	1.1E+09	2.1E+03	2.9E-06	04:32
60.000	1.6E-03	5.5E+08	1.1E+03	1.5E-06	06:48
80.000	1.0E-03	3.5E+08	7.0E+02	9.4E-07	09:05

Table 3: HotSpot Dose Conversion Data for Cs-137

DISTANCE km	T E D E (Sv)	RESPIRABLE TIME-INTEGRATED AIR CONCENTRATION (Bq-sec)/m3	GROUND SURFACE DEPOSITION (kBq/m2)	GROUND SHINE DOSE RATE (Sv/hr)	ARRIVAL TIME (hour:min)
0.030	2.9E-04	1.0E+08	4.3E+01	8.6E-08	<00:01
0.100	7.2E+00	2.5E+12	4.9E+06	9.8E-03	<00:01
0.200	5.0E+00	1.7E+12	3.5E+06	6.9E-03	00:01
0.300	2.7E+00	9.4E+11	1.9E+06	3.8E-03	00:02
0.400	1.6E+00	5.7E+11	1.1E+06	2.3E-03	00:02
0.500	1.1E+00	3.8E+11	7.5E+05	1.5E-03	00:03
0.600	7.7E-01	2.7E+11	5.3E+05	1.1E-03	00:04
0.700	5.8E-01	2.0E+11	4.0E+05	7.9E-04	00:04
0.800	4.5E-01	1.5E+11	3.1E+05	6.1E-04	00:05
0.900	3.6E-01	1.2E+11	2.4E+05	4.9E-04	00:06
1.000	2.9E-01	1.0E+11	2.0E+05	4.0E-04	00:06
2.000	7.6E-02	2.6E+10	5.3E+04	1.0E-04	00:13
4.000	2.1E-02	7.1E+09	1.4E+04	2.8E-05	00:27
6.000	9.7E-03	3.4E+09	6.7E+03	1.3E-05	00:40
8.000	5.8E-03	2.0E+09	4.0E+03	8.0E-06	00:54
10.000	3.9E-03	1.3E+09	2.7E+03	5.4E-06	01:08
20.000	1.2E-03	4.1E+08	8.2E+02	1.6E-06	02:16
40.000	3.8E-04	1.3E+08	2.6E+02	5.3E-07	04:32
60.000	2.0E-04	6.9E+07	1.4E+02	2.8E-07	06:48
80.000	1.3E-04	4.4E+07	8.8E+01	1.8E-07	09:05

 DOI: 10.1051/epjconf/201715600009

From the simulation, total effective dose equivalent (TEDE) and dose rate for Cs-137 and I-131 is calculated for 60 minutes from the onset of the postulated accident. The highest expected TEDE is when the radionuclides blasted out from the NPP core with the value of 60 Sv; with dose rate 0.54 mSv hr^{-1} and 9.8 mSv hr^{-1} for Cs-137 and I-131 respectively. As the dose decreases with time, the detected TEDE after 1 hour is 33 mSv and 3.9 mSv for Cs-137 and I-131 at 10 km from the point source with the respective dose rate of 5.4×10^{-6} Sv hr^{-1} for Cs-137 and 3.0×10^{-5} Sv hr^{-1} for I-131. After 4 hours, the TEDE detected is 0.38 mSv for Cs-137 and 3.2 mSv for I-131, at 40 km distance from the site with dose rate of 5.3×10^{-4} mSv hr^{-1} and 2.9×10^{-3} mSv hr^{-1} respectively.

6. Discussion

This study shows the radioactive dispersion analysis of Cs-137 and I-131 in CS1, should the similar accident equivalent to Fukushima happen in the candidate site. The result shows that both radionuclides can reach the nearest town, Lumut and Sitiawan, which located 18 km from the CS1 in less than 3 hours from the accident time. Both radionuclides can also reach major towns such as Taiping, Ipoh, Kampar and Teluk Intan that located around 60 km from CS1 in less than 8 hours from the accident time. From the observation, although Lumut and Sitiawan are close to the site, both towns are protected by hilly terrains more than 100 meters high, which become a natural barrier from direct radioactive dispersion. Hence, both towns are considered protected from accidental or normal radioactive dispersion of NPP in CS1.

To estimate radiological risks from the postulated accident for Taiping, Ipoh, Kampar and Teluk Intan, total effective dose received from Cs-137 and I-131 during accident is calculated. For Cs-137, estimated total dose received at the major towns which is around 65 km from CS1 is 0.2 mSv (dose rate received less than 7 hours after incident time) during the incident. Since the TEDE and dose rate will decrease with time and distance, the most likely effective dose rate is less than 1 mSv yr^{-1}. For I-131, the estimated total dose rate received at Taiping, Ipoh, Kampar and Teluk Intan is 1.6 mSv during the accident. However, since I-131 half life is only 8 days, the dose rate is expected to decrease drastically throughout the year and may results annual dose rate less than 1 mSv yr^{-1}.

Since the estimated annual effective dose is lower the global limit, the NPP candidate site in this study can be considered safe for NPP construction. However more detailed analysis should be conducted in risk assessment before the decision to select CS1 as NPP site can be made.

7. CONCLUSION

The dispersion analysis for NPP candidate site CS1 is conducted using HotSpot software based on Fukushima Accident in 2011 and it shows that the radionuclide dispersion can reach as far as 80 km from the site, where the Perak capital, Ipoh is located. However, estimated annual effective dose is not more than 1 mSv limit, which is considered acceptable in the point of view of radiological health risk for human and the environment. However, more studies should be conducted to ensure all radiological risks and hazards are identified before the decision to select CS1 as NPP site can be made.

One of the author (SDS) would like to thank the Ministry of Higher Education (MOHE) Malaysia and Universiti Teknologi Malaysia for providing a GUP grant Q.J130000.2526.12H21 for this research.

REFERENCES

[1] WNA, (2015). Nuclear Share Figures, 2004-2014. World Nuclear Association. Web article from: **http://www.world-nuclear.org/info/Facts-and-Figures/Nuclear-generation-by-country/** [Extracted at 19th January 2016].

[2] Jewell, J., (2011). Ready for nuclear energy?: An assessment of capacities and motivations for launching new national nuclear power programs. *Energy Policy*. 39,1041-155.

[3] WNA, (2016). Emerging Nuclear Energy Countries. World Nuclear Association. Web article from: **http://www.world-nuclear.org/info/Country-Profiles/Others/Emerging-Nuclear-Energy-Countries/** [Extracted at 19th January 2016].

[4] Basri, N.A., Ramli, A.T., Aliyu, A.B.S., (2015). Malaysia energy strategy towards sustainability: a panoramic overview of the benefits and challenges. *Renewable and Sustainable Energy Reviews*. 42, 1094-1105.

[5] Aliyu, A.B.S., Ramli, A.T., Salleh, M.A., (2015). Assessment of potential human health and environmental impacts of a nuclear power plant (NPP) based on atmospheric modeling. *Atmosfera*. 28(1), 13-26.

[6] IAEA, (2003). Site Evaluation For Nuclear Installations. International Atomic Energy Agency. Vienna, Austria.

[7] AELB, (2011). Guideline For Site Evaluation For Nuclear Power Plant. Atomic Energy Licensing Board.

[8] Pasquill, F. (1961). The estimation of the dispersion of windborne material. *The Meteorological Magazine*, vol 90, No. 1063, pp 33-49.

[9] Homann, S.G. and Aluzzi, F., (2014). HotSpot Health Physics Codes User's Guide Version 3.0. National Atmospheric Release Advisory Center, Lawrence Livermore National Laboratory, California, 198 pp.

[10] Department of Statistics, (2010). Basic Population Characteristics by Administrative Districts. *Census Report 2010*. Malaysia Department of Statistics. Kuala Lumpur, Malaysia.

[11] Hirose, K. (2012). 2011 Fukushima Dai-ichi nuclear power plant accident: summary of regional radioactive deposition monitoring results. *Journal of Environmental Radioactivity, 111*, 13-17.

[12] Korsakissok, I., Mathieu, A., & Didier, D. (2013). Atmospheric dispersion and ground deposition induced by the Fukushima Nuclear Power Plant accident: A local-scale simulation and sensitivity study. *Atmospheric Environment, 70*, 267-279.

[13] Katata, G., Ota, M., Terada, H., Chino, M., & Nagai, H. (2012). Atmospheric discharge and dispersion of radionuclides during the Fukushima Dai-ichi Nuclear Power Plant accident. Part I: Source term estimation and local-scale atmospheric dispersion in early phase of the accident. *Journal of Environmental Radioactivity, 109*, 103-113.

The influence of annealing temperature and heating rate on thermoluminescence properties of nanocrystalline calcium borate powder

T. N. H. Tengku Kamarul Bahri*, *R.* Hussin , and *N. E.* Ahmad

Department of Physics, Faculty of Science, Universiti Teknologi Malaysia, 81310 Skudai, Johor Bahru, Malaysia

Abstract. We have reported the influence of annealing temperature and heating rate on thermoluminescence (TL) properties of nanocrystalline calcium borate, CaB_2O_4, powder synthesized by solution combustion method. Powder X-ray diffraction experiments were carried out on CaB_2O_4 to get the crystal phase and size. The samples were annealed using the TLD oven and exposed to cobalt-60 source. TL glow curves were measured and recorded using a Harshaw model 3500 TLD reader. The crystal phase confirmed one major phase of CaB_2O_4 with 27 nm in size. CaB_2O_4 has a simple glow curve with only one and a well defined peak at around 150 °C. TL intensity was higher after annealing the material before irradiation which indicated the importance of annealing. It was found that an annealing temperature at 300 °C for one hour and the heating rate of 10 °C s^{-1} was the best procedure to produce high TL intensity.

1 Introduction

Thermoluminescent (TL) dosimetry is used for radiation protection purposes and clinical uses such as for measuring absorbed dose to patient in diagnostic radiology and radiotherapy. Currently, in Malaysia there are 25 radiotherapy centres comprising 7 government hospitals and 18 private medical centres, serving a population of 31.7 million people in 2016. As a member of a state of the International Atomic Energy Agency (IAEA), Malaysia has participated in a project called "IAEA/WHO Thermoluminesence Dosimeter (TLD) Postal Audit Program" [1]. The objective was to provide a quality audit on the dose delivered by megavoltage X-ray teletherapy units for radiotherapy centres worldwide. In Malaysia, total numbers of cobalt-60 and high energy photon beams that have been audited within these years were 19 and 13 units, respectively. TLD-100 (LiF: Mg, Ti) powder was used as a transfer detector in the IAEA and WHO TLD postal dose quality audit because of its good physical aspects including size and tissue equivalent composition and eases of transportation. However, TLD-100 has several notable drawbacks such as being hygroscopic [2]. Also, challenges and requirements of thermoluminescent materials were low cost production and a simple glow curve structure which eliminates the need for complicated annealing regimes as TLD-100 has a complicated glow curve [3]. With these restrictions in mind and the increasing use of radiation in the processes associated with medical and industrial applications have motivated research on new materials with adequate dosimetric properties.

Nanocrystalline from borate compound has revealed that they include some outstanding characteristics such as high sensitivity and applicable to a linear dose response over a wide range of absorbed dose [4-7]. Calcium borate, CaB_2O_4 is one of the widely studied oxide materials because of its excellent structural and optical properties. This compound with a bone effective atomic number (Z_{eff} ~ 12.5) can be counted as a promising material for the future of personal and environmental applications [8]. Gowda et al. [9] had mentioned in his report that radiation interactive characteristics of mixture or compounds that were similar to bone, soft tissue or any other body constituents can be identified for dosimetric purposes. CaB_2O_4 polycrystalline present a good chemical stability due to their very low hygroscopic nature [10] and linear response with X-ray dose ranges 1 to 1000 mGy and 1 to 10,000 mGy, respectively [11]. The solution combustion method is the among the most popular techniques of production nanocrystals, simple and rapid process. The dissolution process enables the amount of each component to be controlled accurately and uniformly mixed in liquid phase, while the combustion part is a time and energy saving method especially for the preparation of complex oxides. Also, the crystal powder of the solution combustion method is mostly homogeneous, more pure and low cost production than the crystal powder obtained via other conventional solid-state methods [12]. The aim of the study herein was to determine the influence of annealing temperature and heating rate on TL properties based on nanocrystalline CaB_2O_4 powder prepared by solution combustion method.

* Corresponding author: tnhidayah2@gmail.com

2 Experimental

2.1 Synthesis

CaB_2O_4 was synthesized using the solution combustion method taking into consideration the following chemical reactions:

$$Ca(NO_3)_2.4H_2O + 4\ H_3BO_3 + 7\ CH_4N_2O + 7\ NH_4NO_3 : CaB_2O_4$$

The starting raw materials used include various proportions of analytical pure grade calcium nitrate, tetrahydrate, ($Ca(NO_3)_2.4H_2O$, 99 %, Merck KGaA, Germany), boric acid (H_3BO_3, 99.5 %, Merck KGaA, Germany), urea (CH_4N_2O, 98 %, Sigma-Aldrich, Missouri) and ammonium nitrate (NH_4NO_3, 95 %, Merck KGaA, Germany). CH_4N_2O functioned as fuel for combustion as it was the most popular and attractive fuels for producing highly uniform, complex oxide ceramic powders with precisely controlled stoichiometry [13]. Meanwhile, NH_4NO_3 was used as oxidizer because of the highly porous nanocrystalline powder obtained. Porous structure played a key role in decreasing agglomeration, increasing surface area and improving sintering behavior [14]. The raw materials were weighed according to the chemical composition of CaB_2O_4 and put in a 100 ml beaker. 15 ml of deionized water was added and thoroughly mixed using a magnetic stirrer for 20 minutes without heating to obtain a uniform solution. The solutions were then poured into porcelain crucible and placed in a muffle furnace pre-heated at 500 °C. Initially, the solution boiled and underwent dehydration, followed by decomposition releasing large amounts of gases (oxides of carbon, nitrogen and ammonia). Then, spontaneous ignition and smoldering occurred which gradually led to an explosion with enormous swelling. The entire combustion process was in 15 minutes. When taken out of the muffle furnace and cooled down, white voluminous foamy porous was obtained, which was easily grinded using mortar and pestle to obtain the synthesized fine, white powders CaB_2O_4. Following the combustion, to remove the deposited carbon and the unreacted organic residues and also to enhance crytallinity of pure compound CaB_2O_4, the heating process was carried out in temperature of 900 °C for a fixed time of one hour. Thus, the as-synthesized powders were placed into alumina crucible and heated in a furnace with temperature 900 °C. Then, the powder samples were taken out of the furnace and left to cool down at room temperature.

2.2 Characterization

In order to check the formation of the samples, X-ray diffraction (XRD) analysis was carried out. The powder samples were scanned by mean of X-ray diffraction using Siemens Diffractometer D5000, with CuKα radiation operating at 40 kV, 30 mA and use Bragg–Brentano geometry at room temperature, 25°C. The diffraction pattern was measured in steps 0.05° with 1s counting time per step, for 2-theta angle (2θ) ranging from 10° to 80°. In this study, Match! software was used for phase identification from powder diffraction data. It compares the diffraction pattern of the sample obtained to a database containing reference patterns in order to identify the phases which are present. The crystalline size, d_{XRD} can be determined from the most intense and sharp diffraction peak by using Debye-Scherrer formula, with assuming that the particles are stress free [15], as given below:

$$d_{XRD} = \frac{0.9\lambda}{\beta\cos\theta} \tag{1}$$

where d_{XRD} is average crystalline size, λ is incident wavelength, θ is the Bragg angle and β is the diffracted full width at half maximum (in radians), caused by crystals.

Before studying the TL properties of samples, the nanocrystalline CaB_2O_4 powders were initially annealed using the TLD oven type LAB-01/400. This annealing procedure consists of changing the temperature, step by step, at a constant annealing time and heating rate. The temperatures used were 100 °C, 200 °C, 300 °C and 400 °C. The annealing time was kept constant at one hour for each temperature with heating rate 10 °C s^{-1}. After annealing, the sample powders were loaded into small black plastic capsules. This capsule was made of opaque polyethylene capsules (IAEA type) of 3 mm diameter, 15 mm length and with 1 mm thick walls. Each capsule contains about 100 mg of TLD-powder. These capsules were then kept in the closed cabinet to avoid direct sunlight and ultraviolet radiation that may contribute to the development of background signal. Temperature within the TLD storage room also was controlled for not higher than 30 °C to maintain the stability of the parameters.

Irradiation was performed using a cobalt-60 source (Gammacell 220 Excel, MDS Nordion, Ottawa, Canada) at dose 100 Gy at room temperature. The set up for beam field size and standard source-surface distance (SSD) that were used was 10 x 10 cm^2 and 100 cm, respectively. After 24 hours irradiation, powder samples were read using a Harshaw model 3500 TLD reader with the Windows Radiation Evaluation and Management System (WinREMS) software. Powder dispenser was used to aid in dispensing accurate amounts of dosimetric powders, taking consistent delivery of a set amount 15 mg to the planchet of the 3500 reader. The process was performed with care to reduce any loss of powder during dispensing.

To read TL material, a time temperature profile (TTP) must be set up. TTP is the temperature to which the TL material is heated as a function of time. It is defined in three segments: Preheat, Acquisition, and Anneal. Preheat establishes a common starting point for all dosimeters in a group and may be used to eliminate the fast-fading low temperature peaks. Acquisition is the segment during which the dosimetric data are acquired and the glow curve is generated. The Anneal segment is used to hold the dosimeter at a high temperature to ensure that the entire TL signal is removed. Data obtained transmits to WinREMS in the form of 200 response points forming a glow curve. The glow curve generated plots the counts versus temperature automatically and the calculation formula in Eq. (2b) needs to be set in Excel in order to convert from counts to nanoamperes.

$$Q(C)/t(s)=I(A) \tag{2a}$$

This can be written as

$$\frac{\text{counts}}{16000\text{ counts}}(nC)\Big/\frac{t_{acq}(s)}{200\text{ glow curve acquisition points}}=I(nA) \tag{2b}$$

where Q is the electrical charge converted from the light captured by the photomultiplier (PMT) that is collected from each sample in 16000 data points along. The t is the time when the dosimetric data is acquired in 200 channels and I is the electric current produced. The acquisition time in seconds (t_{acq}) can be found on the TTP that was created. The total amount of electric current produced and area integral under the glow curves yields the TLD reading (or TL response) signifies the radiation energy deposited and used for measuring the TL intensities. Usually, in units of micro coulombs since current multiplied by time equals charge and coulombs is the unit of charge. In this study, an average TL signal from a set of six readings was calculated and were normalized to unit mass (g) to obtain a net TL signal. The results had to be normalized in order to compensate differences of masses shapes and size of the samples.

3 Results and discussion

3.1 Crystalline phase and size

Fig. 1 shows the XRD pattern of CaB_2O_4 after heat treatment at temperature 900 °C for 1 hour. The major crystalline phase of CaB_2O_4 is formed as the products of the nitridation reaction. This is in good agreement with the reference database from the Joint Committee on Powder Diffraction Standards (JCPDF) card number 75-0640. The obtained product has orthorhombic lattice with the lattice parameters a = 6.19 Å, b= 11.60 Å, and c= 4.28 Å. XRD pattern reported by Huang et al. [16] also indicated the pure phase of CaB_2O_4 at 900 °C. Meanwhile, Yang et al. [17] reported that heating crystallized within the temperature region of 800 °C to 900 °C and mostly in two stages, corresponding to calcium borate and calcium silicate crystallization. The crystalline size was estimated from most intense and sharp diffraction peak at 29.6° by using Eq. (1), which was approximately 27 nm in size.

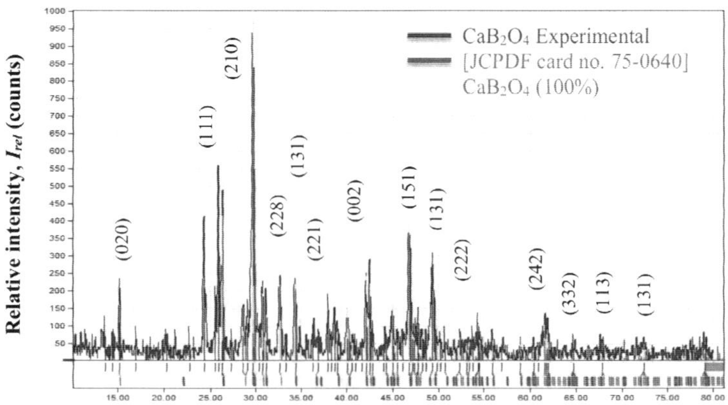

Fig. 1. XRD patterns of CaB_2O_4 after heat treatment at temperature 900 °C for 1 h.

3.2 Effect of annealing

The optimum annealing procedures were investigated in order to determine the highest TL yield sensitivity of the prepared samples. Fig. 2 illustrates the TL intensity of nanocrystalline CaB_2O_4 powder before and after annealing at 300 °C and exposed to 100 Gy of cobalt-60 irradiation. The intensity of nanocrystalline CaB_2O_4 powder after annealing process increases as compared to nanocrystalline CaB_2O_4 powder before annealing process from 1100 nA g^{-1} to 11000 nA g^{-1} at about 90 %. Thus, the annealing process is needed in every process of TLD used in order to increase accuracy and sensitivities of TL signal towards ionizing radiation.

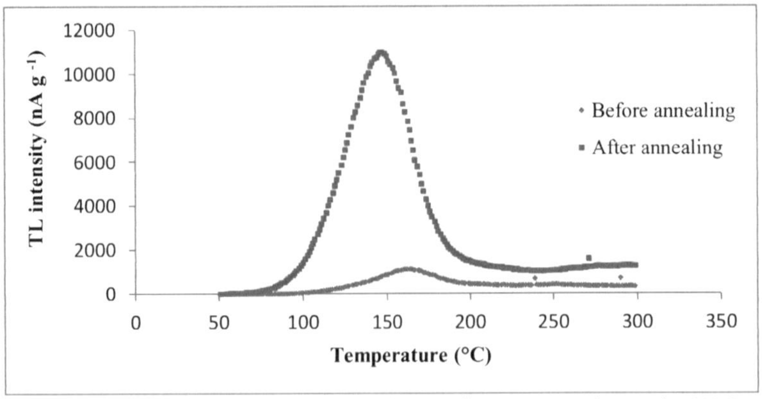

Fig. 2. TL intensity of nanocrystalline CaB_2O_4 powder before and after annealing at 300 °C and exposed to 100 Gy of cobalt-60 irradiation.

Some thermoluminescent material required a complex annealing procedure. Whether it requires a high temperature anneal or a low temperature anneal or both of them. The high temperature anneal is needed to clear the dosimetric traps of residual signal which may cause unwanted backgrounds during subsequent use of the dosimeters. Meanwhile, low temperature anneal is to stabilize and aggregate low temperature traps in order to enhance the sensitivity of the main dosimetry traps and to reduce losses of radiation induced signal due to thermal or optical fading during use [18].

Fig. 3 shows the TL intensity of nanocrystalline CaB_2O_4 powder for different annealing temperature at 1 hour. It displays a simple glow curve for all samples where these samples exhibit only one peak at around 150 °C. The highest TL intensity was detected at 300 °C followed by 200 °C, 400 °C and 100 °C. Fig. 4 indicates the TL response and standard deviation as a function of annealing temperatures from 100 to 400 °C, which corresponds to data analysis from glow curves in Fig. 3. It can be seen that the highest TL response was obtained at 300 °C with the lowest standard deviation approximately 5 %. This means that, the annealing temperature of 300 °C is well sufficient to maximize the emptying of shallow traps and to boost electrons out of the traps. Since increased temperatures in the oven, resulting in increases thermal vibrations of the molecules and the rate of electron release. However, temperature above 300 °C causes irreversible damage to the material. Thus, changes in the molecular structure of the CaB_2O_4 nanocrystalline and reduce or induces changes to the number of electron traps. This would result in a loss of TL sensitivity as shown for annealing temperature at 400 °C. Therefore, among the prepared samples, the sample at 300 °C was the best annealing temperature to produce the efficient thermoluminescence properties. Also, it is recommended that the maximum readout temperature in TLD reader is the same as the annealing temperature as the material become damaged when heated to temperatures above 300 °C.

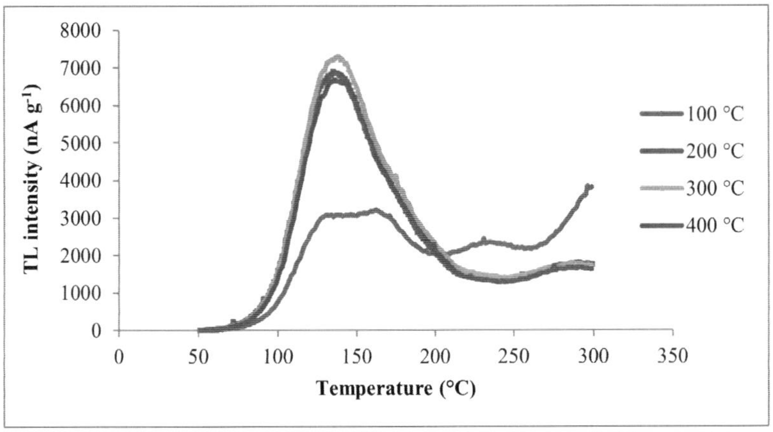

Fig. 3. TL intensity of nanocrystalline CaB_2O_4 powder under different annealing temperature for 1 hour after being exposed to 100 Gy of cobalt-60 irradiation.

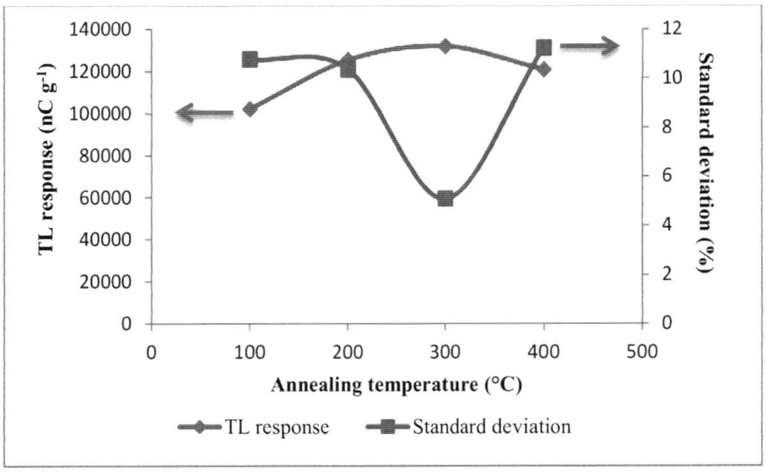

Fig. 4. TL response and the corresponding standard deviation as a function of annealing temperature at 1 hour for nanocrystalline CaB_2O_4 powder.

3.3 Effect of heating rate

The basic requirements for readout of thermoluminescence dosimeters (TLDs) is the heating rate in studying the dosimetric parameters. Fig. 5 shows the TL intensity of nanocrystalline CaB_2O_4 powder recorded at different linear heating rates between 1 °C s^{-1} and 25 °C s^{-1} for 100 Gy of cobalt-60 exposure. As shown in the Fig. 5, as the heating rate increases, the intensity increases and quickly decrease when the heating rate greater or equal to 20 °C s^{-1} up to 25 °C s^{-1}. A thermal quenching effect may be dominating in this portion of the curves where the heating rate $\geq$ 20 °C s^{-1} because the higher the heating rate, the less the probability that clusters will be formed [19]. The implication of formation of trap clusters instead of randomly distributed defects is that it may change the kinetics of trapping and recombination processes, thereby influencing the TL properties of a phosphor [20]. Because of the complex nature of the changes in the cluster formation with heating rate, hence, at some heating rate the cluster formation may be such that enhancement of thermoluminescence was greater than the total effect of reduction, suppression and thermal quenching effect on TL, while at the other heating rates it is vice versa. Also, in case of thermal quenching was present in the temperature range of a composite TL peak, a distortion may appear in the shape of the peak due to progressive quenching as the temperature increases during the heating [21]. Hence, the heating rate was one of the most important experimental parameters that adversely affects the intensities of glow peak of nanocrystalline CaB_2O_4 powder.

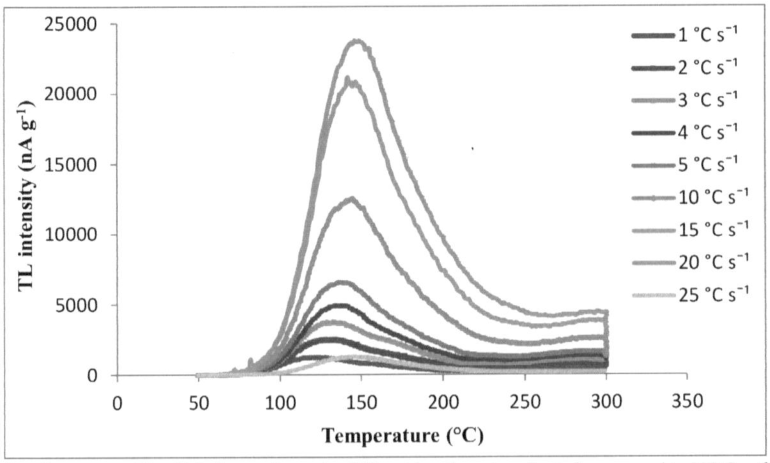

Fig. 5. TL intensity of nanocrystalline CaB_2O_4 powder under different heating rates after being exposed to 100 Gy of cobalt-60 irradiation.

Another important and most interesting result was observed in the behavior of the maximum glow peak temperature. Fig. 6 depicts the maximum glow peak temperature as a function of heating rate of the nanocrystalline CaB_2O_4 powder. There was a shift of the temperature at the maximum intensity of the glow peak (T_{max}) from low temperatures 121 °C to high temperatures 145 °C at heating rate 1 °C s^{-1} to 10 °C s^{-1}. However, T_{max} start to fluctuate at a heating rate greater than 10 °C s^{-1} as shown in Fig. 6. As a result, the peak temperature was 142 °C, 150 °C and 146 °C for heating rate 15 °C s^{-1}, 20 °C s^{-1} and 25 °C s^{-1}, respectively. Thermal fluctuation would allow repeated association and dissociation of such aggregates, a fraction of which might be capable of dissipating the stored energy reserves. If the fluctuations were not present, thus making TL intensity a function of heating rate [22]. A shift in T_{max} with the heating rate has an empirical explanation as proposed by Kitis et al. [23]. At a lower heating rate (β_1), the time spent by the phosphor at a temperature (T_1) is long enough, so that the amount of the thermal release of electrons depending on the half-life at this temperature could take place. As the heating rate increases to $\beta_2 > \beta_1$ the time spent at the same temperature T_1 decreases and therefore the thermal release of electrons is also decreased. Hence a higher temperature T_2 is needed for the same amount of the thermal release to take place at β_2. In this way, the whole glow-peak is shifted to higher temperatures as the heating rate increases in a manner depending on the half-life and the time spent at each temperature. By contrast, a shift of T_{max} toward lower temperatures is due to thermal quenching studied previously by Akselrod et al. [24]. Therefore, there is a slight drop in T_{max} for 15 °C s^{-1} which shows the presence of the thermal quenching effect of thermoluminescence of nanocrystalline CaB_2O_4 powder. This property should be taken into consideration in TL investigations. Hereby, the selection of heating rate of 10 °C s^{-1} has been made for obtaining the highest TL intensity based on the present results of TL intensity and maximum glow peak temperature for different heating rate.

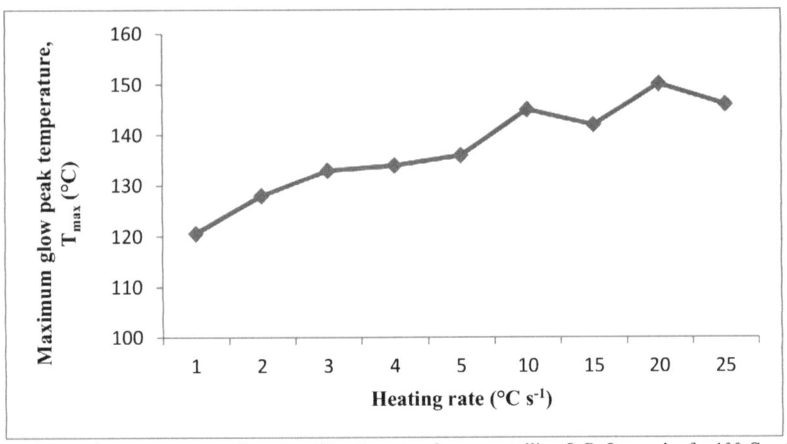

Fig. 6. Maximum glow peak temperature as a function of heating rate of nanocrystalline CaB_2O_4 powder for 100 Gy of cobalt-60 exposure.

The time temperature profile (TTP) in Table 1 lists the best read out parameters for nanocrystalline CaB_2O_4 powder. The heating rate that used was 10 °C s^{-1}. This means that temperatures rise at 10 °C starting from 50 °C for each second. Thus, 26.67 s was needed to reach the maximum temperature, which was 300 °C. Preheat time was 0 s due to no low temperature peak detected to be eliminated. Meanwhile, for anneal time was 0 s because the annealing procedure only carried out in TLD oven. It must be noted that the annealing procedures can be carried out in the TLD reader itself. However, its efficiency is very low when high dose values are involved. The in-reader annealing procedure should be used only if the dose received by the dosimeter is lower than 10 mGy to 20 mGy [25]. Hence, for this reason, the annealing procedure for nanocrystalline CaB_2O_4 powder must be performed in an oven.

Table 1. Time temperature profile (TTP).

Parameter	Value
Preheat temperature	50 °C
Preheat time	0 s
Heating rate	10 °C s^{-1}
Maximum temperature	300 °C
Acquire time	26.67 s
Anneal temperature	300 °C
Anneal time	0 s

4 Conclusion

Nanocrystalline CaB_2O_4 powder was successfully prepared using the solution combustion method. The XRD result confirmed the CaB_2O_4 phase with particle size at about 27 nm by using Debye-Scherrer formula. CaB_2O_4 has a simple glow curve with only one and a well defined peak at around 150 °C. TL intensity was higher after annealing the material before irradiation which indicated the importance of annealing. Annealing temperature and heating rate profiles strongly influenced TL properties of nanocrystalline CaB_2O_4 powder. Annealing temperature above 300 °C causes damage to the material which result in a loss of TL sensitivity. A shift of T_{max} from high temperatures to lower temperatures suggests the presence of thermal quenching at a heating rate greater than 10 °C s^{-1}. As a result of this study, it is recommended to use an annealing temperature at 300 °C for one hour and the heating rate of 10 °C s^{-1} to produce high TL intensity. In order to have a tenable explanation for the variations it will require a more detailed experiment, however, the variations will be due to interplay of more than one phenomenon. This is because a single model may not be sufficient to explain the observed variations. The detailed measurements will be carried out on nanocrystalline CaB_2O_4 powder in further studies to get significant information about the kinetics involved in the process that is obtained from TL glow curve and serves as a tool for evaluating the various TL parameters like activation energy (E) and frequency factor (s).

Acknowledgements The authors would like to thank the Ministry of Education (MOE), Malaysia and Universiti Teknologi Malaysia for providing research grants and facilities. Our appreciation also goes to Mr. Ahmad Takim bin Saring from School of Applied Physics, Faculty of Science and Technology, Universiti Kebangsaan Malaysia, 43600 UKM Bangi, Selangor Darul Ehsan, Malaysia for performing cobalt-60 irradiation.

References

1. J. Izewska, G. Azangwe, P. Bera, SSDL Newsletter, **58**, 17-23 (2010)
2. S. W. S. McKeever and M. Moscovitch, Rad. Prot. Dos., **104**, 263-270 (2003)
3. B. F. Wall, C. M. H. Driscoll, J. C. Strong, E. S. Fisher, Phys. Med. Biol., **27 (8)**, 1023-1034 (1982)
4. V. S. Kortov, Radiat. Meas., **45**, 3-6 (2010)
5. N. Salah, Z. H. Khan, S. S. Habib, Nucl. Instrum. Methods Phys. Res., **269**, 401-404 (2011)
6. L. Singh, V. Chopra, S.P. Lochab, J. Lumin., **131**, 1177–1183 (2011)
7. P. D. Sahare, J. S. Bakare, S. D. Dhole, N. B. Ingale, A. A. Rupasov, J. Lumin., **130**, 258–265 (2010)
8. M. E. Haghiri, E. Saion, N. Soltani, W. S. Wan Abdullah, M. Navasery, K. R. E. Saraee, N. Deyhimi, J. Alloys Compd., **582**, 392–397 (2014)
9. S. Gowda, S. Krishnaveni, T. Yashoda, T. K. Umesh, R. Gowda, PRAMANA- J. Phys., **63**, 529-541 (2004)
10. M. Santiago, C. Grasseli, E. Caselli, M. Lester, A. Lava, F. Spano, Phys. Status Solidi (A), **185 (2)**, 285-289 (2001)
11. Y. Fujimoto, T. Yanagida, M. Koshimizu, K. Asai, Opt. Mater., **41**, 49–52 (2015)
12. E. Nakazawa, T. Mochida, J. Lumin., **236**, 72-74 (1997)
13. S. T. Aruna, A. S. Mukasyan, Curr. Opin. Solid State Mater. Sci., **12**, 44–50 (2008)
14. J. H. Bai, J. C. Liu, Sci. Sinter., **42**, 133-141 (2010)

15. J. I. Langford, A. J. C. Wilson, J. Appl. Cryst., **11**, 102-113 (1978)
16. J. Huang, L. Zhou, Q. Pang, F. Gong, J. Sun, W. Wang, J. Lumin. , **24(6)**, 363–366 (2009)
17. X. Yang, L. Zhang, X. Chen, X. Sun, G. Yang, X. Guo, H. Yang, C. Gao, Z. Gou, J. Non-Cryst. Solids., **358**, 1171–1179 (2012)
18. C. Furetta, *Handbook of Thermoluminescence*, World Scientific, Singapore, 8–10 (2003)
19. N. A. Spooner, A. D. Franklin, Radiat. Meas. , **35**, 59–66 (2002)
20. A. Mandowski, , J., Swiatek, J. Phys. III France, **7**, 2275–2280 (1997)
21. C. M. Sunta, V. N. Bapat, PACT, **6**, 252 (1982)
22. S. K. Bhattacharjee, Curr. Sci., **84 (11)**, 1419-1427 (2003)
23. G, Kitis, M. Spiropulu, J. Papadopoulos, S. Charalambous, Nucl Instrum Methods Phys Res, **73**, 367-72 (1993)
24. M. S. Akselrod, N. Agersnap Larsen, V. Whitley, S. W. S. McKeever, J. Appl. Phys., **84**, 3364 (1998)

Thermoluminescence study of aluminium oxide doped germanium prepared by combustion synthesis method

Nurul Syazlin Binti Saharin*, Nor Ezzaty Ahmad, Abdul Rahman Tamuri,

Department of Physics, Faculty of Science, Universiti Teknologi Malaysia, 81310 UTM Johor Bahru, Johor, Malaysia.

Abstract. The present paper reports the optimum concentration of germanium (Ge) dopant in aluminium oxide (Al_2O_3) samples prepared by combustion synthesis (CS) method for thermoluminescence (TL) studies. The samples were prepared at various Ge concentration i.e. 1 to 5% mol. The phase formation of un-doped and Ge-doped Al_2O_3 samples was determined using X-ray Diffraction (XRD). The sharp peaks present in the XRD pattern confirms the crystallinity of the samples. The samples were then exposed to 50 Gy Cobalt-60 sources (Gamma cell 220). TL glow curves were measured and recorded using a Harshaw Model 3500 TLD reader. Comparison of TL peaks were observed to obtain the best composition of Ge dopants. A simple glow curves TL peak at around 175°C for all composition samples was observed. It was also found that the composition of aluminium oxide doped with 3.0% of germanium exhibits the highest thermoluminescence (TL) intensity which is 349747.04 (a.u).

Keywords: Thermoluminescence; germanium doping; combustion synthesis

1 Introduction

The application of thermoluminescence (TL) dosimetry in radiation protection has grown steadily due to the worldwide progress of the development of solid thermoluminescent dosimeter (TLD). Various kinds of thermoluminescence (TL) dosimeters have been used in radiation dosimetry. One of the materials that can be considered for TLD was aluminium oxide. This is due to aluminium oxide (α-Al_2O_3) shows an effective number close to that of bone (Z_{eff} = 11.28), it becomes a better choice to be used in medical and environmental dosimetry (Bos, 2001). The TL properties of Al_2O_3 samples of different dopants materials which can enhance TL properties have been studied by several researches for nearly 50 years [1,2]

*Corresponding author: syazlin3@gmail.com

The materials were generally synthesized by sol-gel method [3], ion beam implantation [4], solvent evaporation [5] and combustion synthesis (CS) [6,7,8]. Among this method, combustion synthesis (CS) is an effective and low-cost method for production of various industrially useful materials [9,10]. The solution preparation only requires low processing temperatures and short reaction times (~seconds) making it a quick and easy process [11,12,13]. This process is used directly in the production of high purity and homogeneous ceramic oxide powders. The combustion process is due to an exothermic, redox reaction between nitrate ions and fuel. The large volume of gases released during the reaction promotes a rapid disintegration of the precursor gel, yielding the desired nano-crystalline materials [14].

The presence of impurities in a material is important to the thermoluminescence process. Recently, to have a better performance in the dosimetric purpose, Al_2O_3 is always doped with impurities that induce many different types of trapping centers at which charged particles produce ionizing radiation [15]. Since then, there are a lot of efforts have been directed towards the improvements on its sensitivity via various dopants such as Si, Ti, [16], Mg and Y [17], Cr and Ni [18]. Study by Ravichandran., (2014) [19] of undoped and (2-10mol%) Al doped ZrO_2 nano powder were synthesized using solution combustion method having glycine as a fuel which is pre-heated at approximately 500°C. However, the study only concerned on the optical properties of the samples and not on the thermoluminescence properties. Barros *et al*, (2008), reported the highest sensitivity were found for Al_2O_3 samples doped with 5% Tb, 5% Eu, and 5% Si. The samples were exposed to 1.5Gy Co-60 gamma rays. It stated that the TL response were 5000 times higher than the un-doped Al_2O_3. They also reported that a well defined glow curve peaks was observed at approximately 200°C which is well suited for radiation dosimetry.

However, thermolumiescence properties of Ge doped Al_2O_3 prepared by CS method have not been reported so far. Recently, interest of Ge as doped for dosimetric purpose has increased based from previous study Ge doped optical fibre as a thermoluminescence dosimeter in radiotherapy is a promising material because it gives good response and full fill the criteria of thermoluminescence properties [20]. However, the preparation of optical fibre is not simple as CS technique [21]. Hence, the aim of this present study is to prepare undoped and germanium doped aluminium oxide (Al_2O_3: Ge) nanocryatalline powder synthesis by CS method at various concentrations (1-5mol%). The optimum Ge concentration is chosen for TL study.

2 Materials and methods

Al_2O_3 doped with Germanium was synthesized using the combustion synthesis method. The samples were prepared using aluminium nitrate (Al $(NO_3)_3.9H_2O$) as an oxidizer and urea (CH_4N_2O, 98%) as a fuel, by Sigma Aldrich and Germanium dopant were added with concentration between 1 to 5 mol %. Urea is suitable fuels because they are amino acid that can act as complexings agent of the metal ion in the solution and also serves as fuel to the synthesis nanocrystalline metal oxides. The starting materials taken by mixing 15 g of aluminium nitrate, $Al(NO_3)_3$, 6 g of urea, $CO(NH_2)_2$ and different weight dopant concentration from 1% to 5%. Then, 20 mL distilled water was added to the samples and stirred for 15 minutes using magnetic stirrer in order to obtain a homogeneous solution for completely dissolved. Then the solution was poured into alumina crucible and transferred to muffle-furnace pre-heated at 500°C. The sample ignited spontaneously with few seconds and produced white foamy product. The sample were then taken out from the muffle-furnace and allowed to cool down before grinding using mortar. X- Ray diffractometer (XRD) was used to identify the crystalline phase of the samples. Before the samples were exposed to gamma radiation pre-annealing process were carried out. This procedure was done to remove any effects due to any unwanted previous irradiation without damaging the material. The powders were annealed at 400°C for 1h followed by cooling process at 80°C for 16 h [22]. The prepared samples were then irradiated at 50 Gy sources of Cobalt-60 gamma Cell Model 220 Excel at Universiti Kebangsaan Malaysia, Bangi, Malaysia. The TL response from the glow curve was measured using a TLD reader Harshaw Model 3500 with optical detection of light with wavelength between 400 and 430 nm.

3 Results and discussion

3.1 X-Ray Analysis (XRD)

Figure 1 represents the XRD pattern of the as synthesized Al_2O_3 and Al_2O_3 doped with germanium at 3% mol concentration. XRD was done to confirm the phase of the sample either amorphous or crystalline. The diffraction pattern depicted intense peaks thereby, gives a strong indication that the prepared sample has high degree of crystallinity. The peaks found match with the International Centre for Diffraction Data (ICDD) which are Al_2O_3 (01-070-5679) and Al_2O_3: Ge (01-088-2013). Pure Al_2O_3 sample was observed and no other impurity phase was detected. It is to be noted that small amounts of dopants do not produce significant effect on the basic crystal structure of Al_2O_3. The addition of dopant will not change the unit cell parameters must not change [23].

3.2 TL intensity

Figure 2 represents the TL intensity of Al_2O_3: Ge with different concentration (1 to 5mol%). In the same Figure, the standard deviations are also plotted against the doped concentration. The TL intensity of the sample was found to vary with increasing Ge concentration. The TL intensity increase from 1 mol% Ge to maximum intensity at 3mol% Ge concentration 349747.04 (a.u). Then the intensity decreases with further increase of Ge concentration. This may be due beyond 3 mol% concentration, quenching is bound to occur hence the optimum concentration for Ge only suite for this concentration. Highest TL intensity at 3 mol% Ge is attributed to the recombination of the excited electrons from the valences band with defects produced via irradiation [24]. As a conclusion, it is found that the optimum concentration value of Ge dopants is at 3 mol% with very low standard error.

DOI: 10.1051/epjconf/201715600011

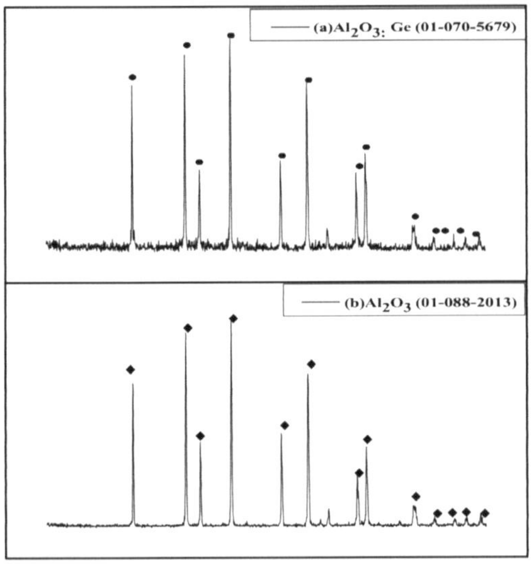

Fig1. XRD pattern of (a) Al_2O_3 doped 3 mol % Ge samples and (b) un-doped Al_2O_3

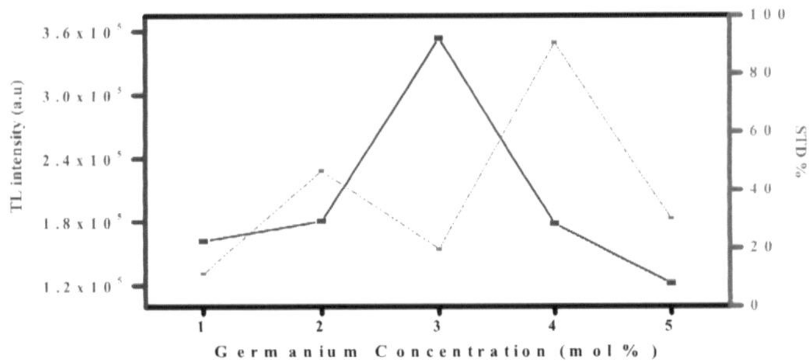

Fig2. TL intensity and standard error of Al_2O_3: Ge with different concentration

3.3 TL Glow curve

Figure 3 displays the glow curve of Al_2O_3 doped with Ge sample at different concentration of germanium. Two peaks are observed, one prominent peak at around 175 °C and another hump at around 230 °C. From the Figure 4, it is observed that the intensity of TL peaks increases with the increase in Ge composition where the highest peak found is for 3 mol% Ge concentration. The result is similar with the work reported by Barros et al. (2008) for Al_2O_3 prepared by CS method as mention before. The results also agree with the work done for Al_2O_3 doped with Tb^{3+} and Tm^{3+} where the Tl glow curve peak was observed at 220°C with a linear response in the dose range of 0.1-1 Gy prepared by same method [25]. Study by J.V.Soares, 2014, [26] showed an intense peak at ~190°C and two other with low intensity at 290°C and 350°C. Hence, based from this of the presence of germanium dopants in Al_2O_3 sample enhanced the TL response and this will full fill the major tools to decide on the TL dosimetry potential of the material. The glow curves are particularly important since they are the main indicators of whether a material can be used for TL dosimeter purposes or not. According from Pekpak, 2010, [27] to obtain a good glow curve, a single peak was normally found at approximately 200°C temperature. Peak at low temperature (approximately of 100°C) at below will normally fade quickly and could not yield any information about radiation content.

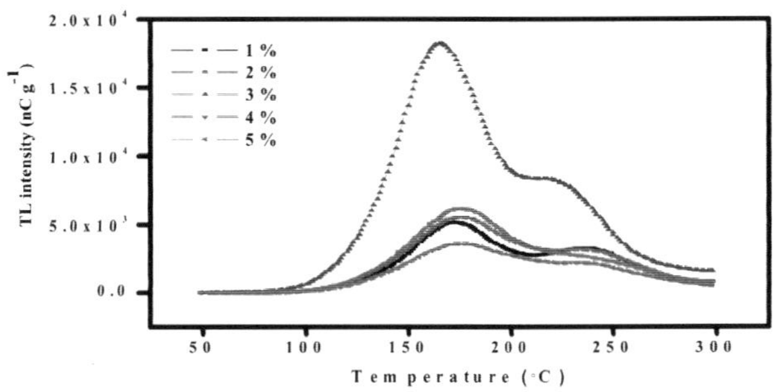

Fig3. TL glow curve of Al_2O_3: Ge at different mole concentration exposed to 50 Gy Cobalt-60

4 Conclusion

From the results above it is possible to conclude that CS method is very suitable for the preparation of Al_2O_3 doped Ge for dosimetric applications. The technique is fast, low cost and produces well defined materials that can be used for dosimetric applications. The concentration analysis shows that the Ge concentration that optimizes the TL response is at 3.0 mol% which has highest TL intensity which is 349747.04 (a.u). The crystallinity of the samples is proven by the XRD sharp peaks. Based from the TL intensity and TL glow curve figure, it can be concluded that the best Ge concentration as Al_2O_3 dopants is 3 mol% where it provides high TL intensity response with low standard error compare other concentration. This value can be used to further study the TL properties.

Acknowledgement

The authors would like to thank the Ministry of Education (MOE) of Malaysia for the financial support (Vote Number: 10H20) and Universiti Kebangsaan Malaysia for providing irradiation source of Cobalt-60 gamma Cell Model 220 Excel and Universiti Teknologi Malaysia for providing research facilities

References

[1] J.F.S., Bitencourt, S.H., Tatumi, Synthesis and thermoluminescence properties of Mg^{2+} doped nano structured aluminium oxide. *Phys. Procedia* 2, 501-514, (2010).

[2] J. K. Rieke, and J. Daniels, Thermoluminescence Studies of Aluminium Oxide. *J. Phys. Chem.* 61(5), 629–633, (1957).

[3] A.A. Kaplyankii, A.B. Kulinkin, A.B. Kutsenko, S.P. Feofilov, R.I. Zakharchenya, T.N. Vasilevskaya, Optical spectra of triply-charged rare-earth ions in polycrystalline corundum. Phys. Sol. State, 40, 1310-1316, (1998).

[4] N. Can, P.D. Townsend, and D.E. Hole, Enhancement of luminescence by pulse laser annealing of ion-implanted europium in sapphire and silica. J. Appl. Phys.78, 6737-6744, (1995).

[5] J. Azorin, Preparation methods of thermoluminescent materials for dosimetric applications: An overview. Appl. Radiat. Isot. 83, 187-191, (2002).

[6] D.V, Barros, S.M. de Azevedo Walter, M. Khoury and Pedro Linhares Filho, Combustion synthesis: A suitable method to prepare Al_2O_3 doped materials for thermoluminescent dosimetry. Radiat. Meas. 43, 345-348, (2008).

[7] G. Hirata, N. Peres, M. Tajeda, J.A. Gonzalez-Ortega, and J. McKittrick, Luminescence study in Eu-doped aluminium oxide phosphors. Opt. Mater. 27,1311-1315, (2005).

[8] V.SM. Barros, W.M.de Azevedo, H.J. Khoury, M.E. A. Andrade, and P. Linhares Filho, Thermoluminescence study of aluminium oxide doped with terbium and thulium. Radiat. Meas. 45,435-437. (2010).

[9] B.N. Lakshminarasappa, J.R. Jayaramaiah, & B.M. Nagabhushana, Thermoluminescence of combustion synthesized yttrium oxide, Powder Technol. 217, 7–10, (2012).

[10] A. Sharma, A. Rani, A. Singh, O.P. Modi, & G.K. Gupta, Synthesis of alumina powder by the urea–glycine–nitrate combustion process: a mixed fuel approach to nanoscale metal oxides, Appl Nanosci. 4(3), 315–323, (2014).

[11] R.K. Lenka, T. Mahata, P.K. Sinha, & A.K Tyagi, Combustion synthesis of gadolinia-doped ceria using glycine and urea fuels, J. Alloys Compd. 466, 326–329. (2008).

[12] R. Garcia, and G.A. Hirata, New combustion synthesis technique for the production of $(In_xGa_1\text{-}x)_2O_3$ powders: Hydrazine/metal nitrate method, 16,1059-1065, (2001).

[13] B.N. Lakshminarasappa, J.R. Jayaramaiah, & B.M. Nagabhushana, Thermoluminescence of combustion synthesized yttrium oxide, Powder Technol. 217, 7–10, (2012).

[14] M.C. Gardey Merino, G.E. Lascalea, L.M. Sánchez, P.G. Vázquez, E.D. Cabanillas, & D.G. Lamas, Nanostructured aluminium oxide powders obtained by aspartic acid-nitrate gel-combustion routes, J. Alloys Compd. 495(2),578–582, (2010).

[15] L.C. Yong, H. Wagiran, A.K. Ismail. Thermoluminescence Performance of Carbon-doped Aluminium Oxide for Dose Measurement by Various Preparation Methods, Jurnal Teknologi. 62:3, 109-113, (2013).

[16] Mehta, S.K., Sengupta, S.,b. Al_2O_3 phosphor for Thermoluminescence dosimetry. Health Phys. 31, 176-177. (1976)

[17] Osvay. Ma.,Tamas BIRO.,1980. Aluminium oxide in TL Dosimetry. Nucl. Instr. Meth, 175, 60-61.

[18] Lapraz, D., P. Iacconi, D. Daviller, and B. Guilhot. 1991. Thermostimulated Luminescence and Fluorescence of Alpha- $Al_2O_3:Cr^{3+}$ Samples (Ruby). Phys. Status Solid (A).126, 521-531.

[19] Ravichandran, A.T., Catherine, K., Pushpa, S., Ravichandran, K., Karthika, K., Nagabhushana, B.M., Mantha, S., and Swaminathan, K. (2014). Effect of Al doping on the structural and optical properties of ZrO_2 nanopowders synthesized using solution combustion method. *Superlattices and Microstructures*.75, 533-542.

[20] N.M. Noor, M. Hussein, T. Kadni, D.A. Bradley, A. Nisbet, Characterizationof Ge-doped optical fibres for MV radiotherapydosimetry.Radiat.Phys.Chem.98, 33–4, (2014).

[21] C. L. Ong., S., Kandaiya, H. T., Kho& M. T. Chong, Segments of a commercial Ge-doped optical fiber as a thermoluminescent dosimeter in radiotherapy. Radiation Measurements. 44, 158-162, (2009)

[22] F.O. Ogundare, S.A. Ogundele, M.L. Chithambo, & M.K. Fasasi, Thermoluminescence characteristics of the main glow peak in a -Al 2 O 3: C exposed to low environmental-like radiation doses, J. Lumin. 139, 143–148, (2013).

[23] S. A.,Pardhi, Nair, G. B., R.,Sharma, & S. J.Dhoble, Investigation of thermoluminescence and electron-vibrational interaction parameters in SrAl 2 O 4 :Eu 2+ , Dy 3+ phosphors. Journal of Luminescence, 187, 492–498. (2017).

[24] S. P Puppalwar, S. J Dhoble, N. S Dhoble, & A. Kumar, Nuclear Instruments and Methods in Physics Research B Luminescence characteristics of Li 2 NaBF 6: Cu phosphor, *274*, 167–171, (2012).

[25] M.E.A. Andrade, W.M. Azevedo, V.S.M. Barros, & H.J. Khoury, Thermoluminescence of aluminum oxide co-doped with terbium and thulium obtained via combustion synthesis, Radiat. Meas. 46(12), 1474–1476, (2011).

[26] J.V. Soares, C.F. Gugliotti, Y.S. Kawashima, S.H. Tatumi, & J.C.R. Mittani, Thermoluminescence and optically stimulated luminescence characteristics of Al_2O_3 doped with Tb, Radiat. Meas. 71, 78–80, (2014).

[27] E., Pekpak, A., Yilmaz, and G., Ozbayoglu, An Overview on Preparation and thermoluminescence Characterization of Lithium Borates for Dosimetric Use. The Open Mineral Processing Journal. 3, 14-24, (2010).

DOI: 10.1051/epjconf/201715600012

Assessment of safety distance between components of nuclear plant and study of the vulnerabiliy of the damage caused by an explosion

Aminu Ismaila [1,3], *Rafiziana* Md Kasmani[2*], *Koh* Meng-Hock [1], and *Ahmad* Termizi Ramli [1]

[1]Physics Department, Universiti Teknologi Malaysia, 81310 Skudai, Johor, Malaysia
[2]Energy Engineering Department Universiti Teknologi Malaysia, 81310 Skudai, Johor, Malaysia
[3]Department of Physics, Ahmadu Bello University Zaria, Nigeria

Abstract. This paper deals with the assessment of external explosion, resulting from accidental release of jet fuel from the large commercial airliner in the nuclear power plant (NPP). The study used three widely prediction methods such as Trinitrotoluene (TNT), multi energy (TNO) and Baker-strehow (BST) to determine the unconfined vapour cloud explosion (UVCE) overpressure within the distances of 100-1400 m from the first impact location. The containment building was taken as the reference position. The fatalities of persons and damage of structures was estimated using probit methodology. Analysis of the results shows that both reactor building and control-room will be highly damaged with risk consequences and probability, depending on the assumed position of the crash. The structures at the radial distance of 600 m may suffer major structural damage with probability ranging from 25 to 100%. The minor structural damage was observed throughout the bounds of the plant complex. The people working within 250 m radius may get affected with different fatality ranging from 28 to 100%. The findings of this study is valuable to evaluate the safety improvement needed on the NPP site and on the risk and consequences associated with the hydrocarbon fuel release/fires due to external hazards.

1 Introduction

External hazards (e.g. aircraft impact, flooding and earthquake) can be a significant contributor to the risk in case of nuclear power plant (NPP) operation and can pose serious hazards to the public and environment due to release of hazardous radiation resulting from fire-induced failures of important plant safety systems [1-3]. It may challenge the available emergency services and affect the mechanism for a safe reactor shutdown and this could led to unsafe condition with the potential to cause for reactor core damage. The risk assessment has been the conventional method for assessing the risk and consequences of fire or any

* Corresponding author: rafiziana@utm.my

form of explosion in the NPP. In the risk assessment method, the consequences of possible accidents and disasters are estimated. The safety standards issued by the regulatory agencies are based on the outcome of fire risk analysis. It appears that the consequences from the occurrence of 'incredible events' were not fully addressed in in the probabilistic risk assessment. This has been strongly underlined by the Fukushima Daiichi NPP incident of March 11, 2011 and World Trade Centre (WTC) aircraft attack of September 11, 2001 [4]. For an accident involving aircraft impacting a nuclear structure, consideration should be given to the hazards of impact loading of the aircraft and jet fuel initiated fires after the incident. These requirements are contained in the IAEA documents (NS-G-1.5 and NS-G-3.1).

The factors which determine the severity of aircraft induced explosion include size and speed of the plane, pulse duration, interaction of subsystems, quantity of fuel involved in the explosion, ignition location and time, turbulence level in the plant, scale of the plant, location and shape of obstacles, blockage ratio, early response to the fight fire and availability of firefighting equipment [5, 6]. Of particular interest to this article is the congestion level and location of the obstacles. These are directly related to spatial arrangement of structures and may have a strong influence on the evolution of fireball, pressure build-up and propagation. Therefore, assessment of safety distances between important NPP components for the postulated aircraft incident is the main focus of this study.

This work aims to measure the explosion overpressure and pulse duration that may be experienced by the reactor containment and control-room walls for the hypothetical aircraft incident. Three empirical models (TNT, TNO and BST) were used to determine the overpressure and pulse duration at a given distance from the point of explosion. Those data (overpressure and pulse duration) were used to assess safety distances between building components within the NPP vicinity. The vulnerability of persons and structures were assessed using probit methodology. The subsequent pool fire and fireball effects were not considered in this paper.

2. Theory

The overpressures generated due to UVCE and explosion's effects are primarily influenced by the degree of confinement, weather conditions, quantity and flammability of fuel as well as source and strength of ignition [7]. The assumptions used in developing the empirical models for the UVCE assessment are generally based on these four parameters. This research work uses the most widely empirical predictive models to estimate the consequences of the blastwave produced by UVCE as a function of distance. A brief description of the models are provided in the following sections.

2.1 Equivalent TNT mass Method (TNT)

This empirical method is based on the assumption that the entire flammable volume of cloud takes part in the explosion and that the explosion is centred in a single location [8]. The explosive energy of TNT served as the benchmark against explosive power of other materials [9]. In the method, the explosive energy of UVCE is expressed as an equivalent mass of tri-nitrotoluene (TNT) that would produce the same explosive energy [10]. In other words, the destruction arises from the mass of TNT is considered to be the same to that of flammable substance under investigation [7]. The explosive potential is primarily

 DOI: 10.1051/epjconf/201715600012

determined by the amount of fuel released. The equivalent-charge-mass of TNT is calculated as in Equation 1.

$$M_{TNT} = \frac{f_E \Delta H_c M_f}{\Delta H_{TNT}} \quad (1)$$

where M_f (kg) is the mass of fuel that involved in the explosion, ΔHc (J/kg) is heat of combustion of the fuel, ΔH_{TNT} (J/kg) is heat of combustion of TNT (4,760 kJ/kg), and f_E is the TNT efficiency factor which depend on the degree of confinement in the area. A value of 10% was used in this study.

For TNT calculation, the Hopkinson-Cranz cube-root scaled distance is given as

$$Z = \frac{x}{M_{TNT}^{\frac{1}{3}}} \quad (2)$$

where Z is the scaled distance (m $kg^{-1/3}$), M_{TNT} is the TNT mass (kg) equivalent and is the distance (m) from the centre of explosion. The corresponding value of overpressure of the shock wave resulting from the UVCE for any given scale distance can also be computed using Equation 3 [10].

$$P_s = \frac{808\left(1+\left(\frac{Z}{4.5}\right)^2\right)}{\sqrt{1+\left(\frac{Z}{0.048}\right)^2}\sqrt{1+\left(\frac{Z}{0.32}\right)^2}\sqrt{1+\left(\frac{Z}{1.35}\right)^2}} \quad (3)$$

The TNT method is simple and easier to use. It is also attractive if the overall damage potential is the only concern particularly for estimating far field damage as reported. However, the biggest problems with this method is that it is not applicable to small distances and overpressures greater than 1 bar. The local pressure evolved in case of TNT is much higher than that of flammable gas explosion waves which travels to a larger distance [8]. A typical error of 20-30% for the explosive performance between TNT and other higher explosives have been reported [11]. Therefore, this method does not estimate overpressure correctly for a far field regions as in most cases it is being over-predicted. Consequently, the predicted blast modelling results could be said to be statistically unreliable. Despite of the drawbacks due to its simplified assumptions, the TNT model is still widely applied in the prediction of blast overpressure at a particular at a particular point of interest from the explosion centre.

2.2 TNO Multi Energy Method (TNO)

This method is based on the basic assumption that detonation can only take place in the obstructed region and could be ruled out in the unconfined parts of the region. The congested regions will contribute to a higher strength of explosion blast. Unobstructed parts of the cloud will slowly burn, without much contribution to the strength of the blast [10-12]. This assumption is supported by vast experimental data [13, 14]. The energy, E (MJ) released by the explosion and Sachs-scaled distance, r' (dimensionless) are given by Equation 4 and 5 respectively.

$$E = \left(V \left[\Delta H_c * \rho * \frac{1}{15.6} \right] \right) \tag{4}$$

where, ΔHc is the heat of combustion of explosive (43 MJ/kg), V is the cloud volume, ρ is the density (810 kg m^{-3}) and the factor ($\frac{1}{15.6}$) is the stoichiometry of the reaction (ratio of fuel: air).

$$r' = x\left(\frac{E}{P_a}\right)^{-\frac{1}{3}} \tag{5}$$

where, x is the distance from the centre of explosion, E is the energy released during the explosion, and P_a is the ambient pressure (0.1 MPa). A step by step on using this method is described [12].

With the equation 4 and 5, the blast overpressure P (MPa) can be calculated as follows:

$$P = P_s' * P_a \tag{6}$$

where P_s' the Sachs-scaled overpressure (dimensionless) and Pa is the ambient atmospheric pressure (MPa). The Sachs-scaled overpressure is estimated from the curves consisting of scaled overpressure versus dimensionless scaled distance.

The time duration of positive phase t_p (s) is given by the equation:

$$t_p = (t_p'/C_s)(E/P_a)^{\frac{1}{3}} \tag{7}$$

where t_p' is the Sachs-scaled pulse duration obtained from the curve consists of positive pulse duration versus scaled distance shown in the researches by [12, 13], C_s is the velocity of sound in air (340 m s^{-1}), E is the energy released (MJ) and P_a is the ambient pressure (0.1 MPa).

It was observed in various studies that empirical results from the TNO multi energy method are generally much closer to the experimental data based upon the most common damages that occur from explosion [8, 14]. This method was use for vulnerability calculations.

The assumption of 54 tonne of jet fuel will produce a vapour cloud of 1030.85 m^3. The radius, R (m) of the cloud being considered as a hemisphere is calculated to be 7.895 m. The volume of the congested region (120*60*20 m) is found to be 144,000 m^3. This will give an empty space of 143,933.92 m^3 for the dispersion of the cloud. The assumption that 30% of the fuel capacity to be involved in the explosion will make this part to generate blast of high strength (i.e, higher than 7). A higher strength explosion blast of 10 is used in this scenario. The energy E release by the explosion is calculated as 2301566.4 MJ

 DOI: 10.1051/epjconf/201715600012

(2301.5664 GJ). The energy released during explosion is very close to resultant total fuel energy adds up (54 tonne x 43 $GJtonne^{-1}$ = 2322 GJ) showing that there is energy conservation.

2.3 Baker-Strehlow-Tang Method (BST)

This technique is similar to multi-energy method only differs in their assumptions. The BST model is based on the assumption that the strength of the blast wave is directly related to the maximum flame speed attained within the cloud. Therefore, the important parameter in the selection of the intensity of the explosion blast in this method is the flame propagation speed [15]. The determination of explosion energy follows the recommendations in the Multi-Energy method. The intensity of UVCE is determined by the following factors: (i) degree of confinement (ii) density of the obstacles (iii) the reactivity of the fuel and (iv) the way the flame front propagates [13]. The procedure from TNO method was adopted in determining the vapour cloud dimension and energy released during explosion. The scaled overpressure is calculated from the curve consisting of scaled overpressure as a function of scaled distance, with the flame speed (in form of March number, Mf) as a parameter described in [11, 12]. The reactivity of jet fuel is higher and blockage ratio of the plant in this scenario is considered to be higher than 40%. A 2D flame expansion at a speed of 1.77 is considered in this study.

3. The Scenario

3.1 Description of the Scenario

A generic site plan of a "hypothetical nuclear power plant" is shown in Figure 1. The plant dimensions is 1 km × 1 km. A reinforced concrete reactor containment composed of a circular base slab, an upright cylindrical walls and torispherical dome is placed between control room and stack. The thickness of inner containment is 1 m and outer containment wall (as well as dome) is 1.2 m in line with IAEA guideline [16]. The inner containment is to provide biological shielding while the outer containment is to protect the reactor from external threats. The height of outer containment building is 62 m and 42 m inner radius. The control building that houses the main control room is located on the central location as the hub of the plant operating staff's activities. The congested area consisting of control building, service building, tank storage and condensate storage tank has an average dimension of 120 m x 60 m x 20 m. The arrangement of the building structures is shown in Figure 1.

A hypothetical scenario of aircraft impacting a nuclear reactor containment is considered. About 54 tonne of jet fuel (kerosene equivalent) is assumed to involve in the explosion immediately after the incident. The site layout and blast radius after explosion is shown in Figure 1.

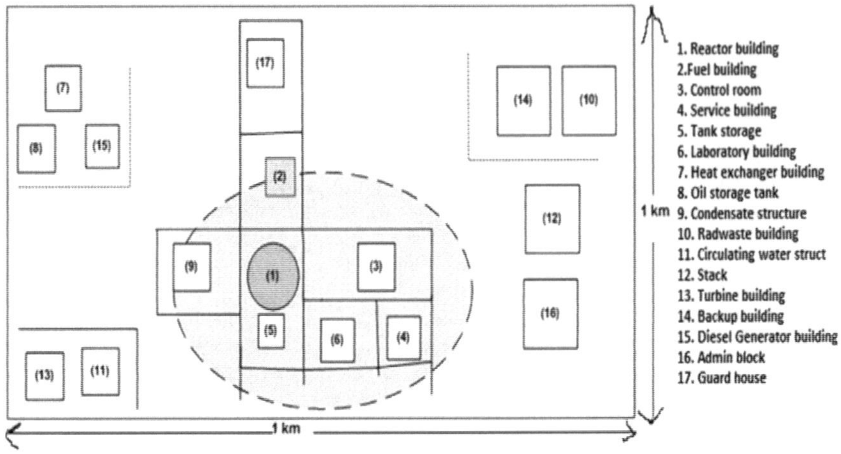

Fig. 1: The plant layout

3.2 Vulnerability Estimation

The vulnerability of individuals and structures to the effects of explosion generally follows probabilistic approach in form of probit (probability unit) methodology. The vulnerability estimates gives an idea of the number of persons who can possibly be affected by explosion in terms of important injuries or casualities. The injuries are that by head impact, eardrum rupture, body displacement or death from lung damage .It also gives an idea on the damage level experienced by the srtuctures as minor or major. It is of interest to note that the probit unit represent an approximation for the probability of surval after impact of the body part.
In this study, probit equation developed by [17, 18] was used to evaluate the direct effects of explosion on structures. The structural damage equation is given by

$$P_r = 5.0 - 0.26 \log_e V \tag{8}$$

where V is the is the measure of the intensity of the damage-causative factor and varies according to the damage intensity.

For major structural damage:

$$V = \left(\frac{17{,}500}{p_s} \right)^{8.4} + \left(\frac{290}{i_s} \right)^{9.3} \tag{9}$$

and for minor structural damage:

$$V = \left(\frac{4{,}600}{P_s} \right)^{3.9} + \left(\frac{110}{i_s} \right)^{5.0} \tag{10}$$

 DOI: 10.1051/epjconf/201715600012

where i_s is the impulse of the shock wave $\left(\frac{1}{2}p_s t_p\right)$, P_s is the peak or maximum overpressure (Pa) and t_p is the positive phase duration (s).

Several probit equations which estimate the fatality of persons from head impact, eardrum fracture, and death from possible lung damage and whole-body displacement have been proposed. An equation for estimating lung damage and death from explosion has been proposed by [18] and is given as

$$P_r = 5.0 - 5.74 \log_e S \tag{11}$$

$$\text{where } S = \frac{4.2}{\bar{P}} + \frac{1.3}{\bar{I}} \tag{12}$$

The parameters $\bar{P}\left(\frac{p}{p_a}\right)$ in equation (12) is the scale overpressure on person depending on orientation and $\bar{I} = \left(\frac{i}{m^{\frac{1}{3}}\sqrt{p_a}}\right)$ is the scale impulse which is a function of mass, m (kg). A vertical orientation for a 75 kg mass of an adult men is used in this study.

The correlation equation for estimating eardrum damage due to explosion is given by Equation 13 [18].

$$P_r = -12.6 - 1.524 \log_e P_s \tag{13}$$

And for head impact is given by Equation 13 [18].

$$P_r = 5.0 - 8.49 \log_e S \tag{14}$$

$$\text{where } S = \frac{2.43 * 10^3}{P_S} + \frac{4 * 10^9}{P_s i_s} \tag{15}$$

A correlation equation for death due to whole body displacement given as follows [18, 19].

$$P_r = 5.0 - 2.44 \log_e S \tag{16}$$

$$\text{where } S = \frac{7.38 * 10^3}{P_s} + \frac{1.3 * 10^9}{P_s i_s} \tag{17}$$

4. Results and Discussion

4.1 Variation of overpressure with distance

Section 3.48 of the IAEA safety standard [20] elaborates that hazards due to chemical explosions should be perform on the site and shall be expresses in terms of overpressure and toxicity (if applicable) taken into account the effect of distance. The site should be consider unsuitable if the assessment of such external induced event could not meet the acceptable safety criteria or practicable solutions. The premise is that the effects of overpressure decreases with the increasing distances and vice versa. Therefore, a reasonable physical separation between structural components of the NPP should be provided at a location where the chemical explosion is envisage. A graph of overpressure against distance provides values of overpressures corresponding to a given distance and hence gives an idea of the degree of damage to structural components as a function of distance to an explosion.

The functional relationship between overpressure and distance from the explosion centre is presented in Figure 2. Analysis of the results show that there is higher blast pressure and long positive phase duration at a distance less than 700 m from the blast starting point. The damage corresponding to these overpressure levels are considered unbearable for most building structures. About 50 to 75% of all outer walls could be heavily damaged by the higher overpressure within these distances. The TNO method predicts higher overpressure as compared to other two methods at the same distances. However, the data for the TNO and BST are somewhat comparable for most data points. The main basic difference for the two models occur only at some few points as shown in the inside graph of figure 2. The TNT method gives the lower values in the data points. The findings of this study agrees closely with that of (Soman and Sundararaj, 2012). However, in the study by (Lobato et al., 2006), it was observed that TNT model predicts higher values of overpressure in contradiction to our findings. The present authors are on the opinion that the variation in the results for the two models may be attributed to differences in the assumptions and scenario used. It was noted that the overpressure values obtained by TNO multi energy method are much closer to the experimental values based upon the damage could occur from explosion reported by (Marc and Konstantinos, 2010). Therefore, values of explosion overpressure obtained on the basis of multi energy approach were used for vulnerability estimation.

In the earlier study [6], it was observed that the structures at a separation distance of 600 m may likely suffer total destruction due to extremely higher overpressure generated from the explosion. At a distance of 800 m, the blast overpressure is 0.08 bar while at 1000 m, it was 0.07 bar. These values are considered unsafe as reported in the experimental results by [21-23]. They claimed that 0.068 bar could cause partial demolition of a typical brick buildings, which become uninhabitable. In other words, about 25% of all walls have failed and serious damage to the load carrying elements occurred. Therefore, taking into account our scenario whereby other building auxiliaries such as laboratory, tank storage, service building and control room are positioned at a distance below 800 m from the containment, it could be said that the structures within the radial distance of 1 km are not safe. When TNO method is applied to the Buncefied and Sitapura explosion incidents [24, 25], which have assumed the volume blockage ratio of 4%, it estimated that the overpressure to be greater than 20 bar at a distance of 300 m, 0.09 bar at a distance of 2 km and window break up to 8 km from the site. For a safety consideration, the overpressure received by the target element should be less than 0.02 bar. This could only lead to a limited minor damage like glass breakage or scabbing of the concrete structure.

Accordingly, at a distance of 1300 m, the overpressure estimated was 0.05 bar. This value is within the range considered to cause minor structural damage. Hence, this could not cause much safety concern when considered the safety measures in a typical NPP. On considering the engineering design of the reactor containment (reinforced concrete structure of thickness > 1 m), the authors are on the opinion that collapse could not likely happen due to that postulated overpressure. However, other building auxiliaries may strongly damage. Therefore, the consequences will be determined by the failure of the safety components within the building and its interaction to other safety systems in the reactor building. This work did not address on what safety component of the plant will be affected by the explosion or fireball and how the failure of one safety system may affect the plant operation. This aspect would be treated by the probabilistic safety analysis. The primary concern of this article is to verify on what should be the minimum safety distance between building structures for a postulated aircraft incident in the NPP vicinity.

Another important safety concern in this scenario is the longer pulse duration as it has direct relationship to blast pressure impulse of the shock wave. At a distance of 100 m from the containment, the pulse duration was 126.96 ms and this translates to a detonation speed of 788 ms^{-1}. Taking into account our scenario, the most important to be taking into consideration is the control room which served as the hub for the plant operating personals situated close to the containment. The explosion overpressure at this position may cause serious damage to this building thereby the manual/automatic control systems would lose their functions. It is therefore appropriate to improved positioning and safety measures of the control building within the bounds of the complex. In this case, the positioning should be at least 1300 m from the containment.

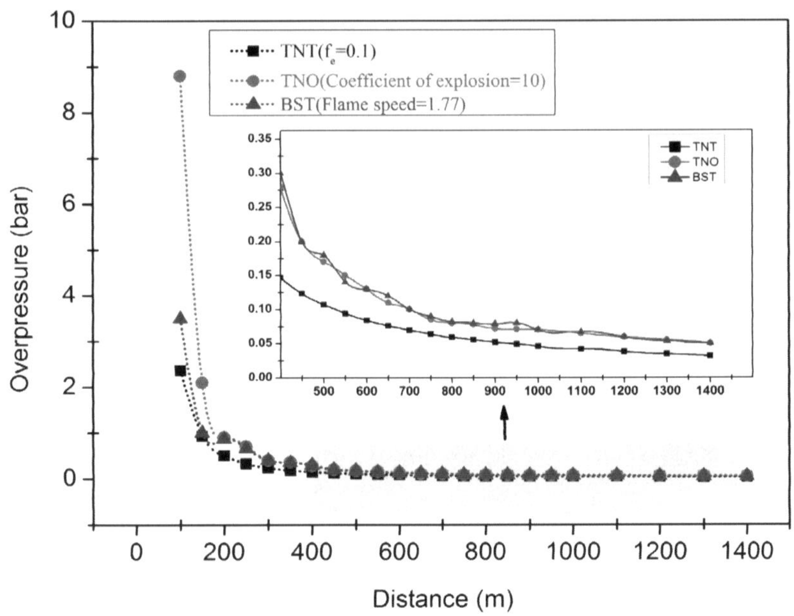

Fig. 2: Overpressure vs distance

4.2 Major and Minor Structural Damage

A major damages are such a level of damage that caused total destruction of buildings which can no longer be restored or useable. By heavy damage, more than 75% of all outer walls have collapsed or majorly damaged and a number of load bearing structural elements have failed and structure has partially collapsed [18]. In case of minor damage, the peak overpressure in the incident wave is expected to cause failure of doors and windows, and that minor crack formation will appear in the load- carrying elements and in the outer walls. Also roofs, ceilings, glass panels and wall covering will be partially destroyed.

Figure 3 presents the percentage of measure and minor structural damage as a function of distance computed using Eq. 8. The vulnerability estimates show that at a distance of 800 m, the probability of major damage to structures is 4.5%. However, it is 46% at a distance of 500 m. Therefore, from the safety point of view, the safety distance from the containment should not be less than 800 m where the probability of major structural damage is less than 5%. Accordingly, the minor structural damage is higher up to a distance of 1000 m with a probability of 66%. This analysis is in agreement with the result obtained by [7] which revealed that a minimum of 1 km safety distance (European Norm) should be considered for higher hazards sites similar to Buncefield. Though, the focus of this article is the assessment of vapour cloud explosion in the NPP site, our analysis showed that the recommended safety distance for hydrocarbon fuels storage depot could also be applied to the NPP site particularly if the blast effect is of prime interest. This assertion is applicable only if the release is less than 55 tonnes. For such worst case scenarios where more than 90 tonnes of jet fuel involved in the explosion, the effects can be reach up to 2 km. In case of Buncefied incident, the effect (windows broken) from overpressure (>20 bar) was observed up to 8 km away [24, 25].

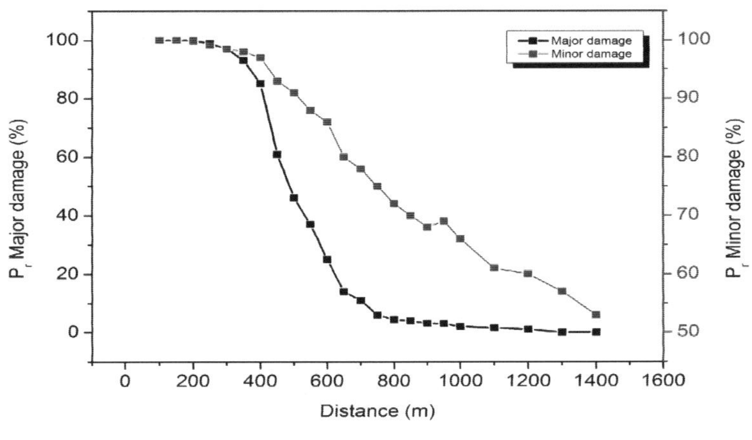

Fig. 3: Major and minor structural damage

4.3 Estimation of Vulnerability of persons

Figure 4 shows the effect of explosion on people for our scenario. For the probability of death due to eardrum rupture, the percentage fatality is calculated as 100 % at a distance of 100 m whereas it is 7% at a location of 300 m. For head impact, the percentage fatality is

DOI: 10.1051/epjconf/201715600012

100 % at a distance below 230 m whereas is zero at a distance greater than 300 m. The percentage fatality due to whole body displacement is estimated as 100% at a distance of 100 m but 3% at 200 m, suggesting that there would be no survival for distance below 100 m. The percentage fatality is zero for distances above 250 m. For death due to lung damage, the percentage fertility is estimated as 100% at a distance of 350 m and above from the centre of explosion but it is zero at 150 m distance. This is quite inconsistent result as the lower distances should experience greater damage. This peculiar condition might be explained as follows; the threshold of lung hemorrhage is 0.68 bar and based on Figure 1, at distance of 200 m, the overpressure is ~ 0.7 bar. The personnel might be escaping themselves once the alarmed went off, resulting to fatality of 100% at distance of 300 m as the overpressure exceeds the threshold.

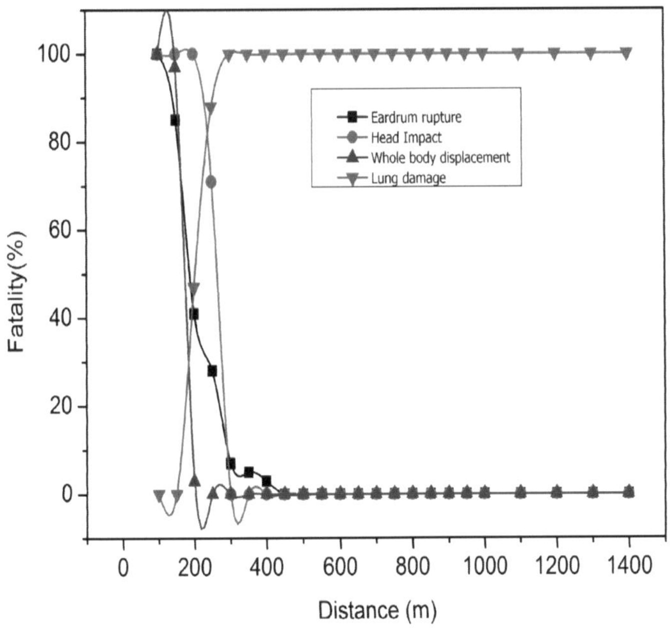

Fig. 4: Vulnerability of persons

5. Conclusion

The consequence analysis of UVCE resulting from the accidental release of 54 tonne of jet fuel in the vicinity of NPP is presented. The three empirical methods for explosion assessments (TNT, TNO multi energy and BST) were used to compute explosion overpressure as a function of scaled distance. At a distance of 1300 km, the Ps values for the TNT, TNO and BST were 0.022 bar, 0.050 bar, and 0.050 bar respectively. These values are within the range causing window breakage and minor structural damage. Taking into account our scenario, some common conclusions drawn from this study are summarised as follows:

(i) The structural components within the bounds of the plant complex are vulnerable to major damage when fuels of 54 tonne is released causing vapour cloud explosion.

(ii) The minimum safety distance from the containment should not be less than 1 km where probability of major structural damage is less than 3 %. The minimum safety distance can go as higher as 2 km if a worst scenario is assumed (i.e. explosion of over 90 tonne).

(iii) The magnitude of explosion overpressure generated from the incident may not likely caused the collapse of reactor containment due to its robust civil engineering design. However, other structural components within the radial distance of 800 m may severely be affected. Further analyses using CFD models are needed to establish the limiting overpressure value that can lead to collapse.

The authors wishes to thank the Ministry of Higher Education Malaysia (MOHE) and Universiti Teknologi Malaysia (UTM) for providing research grants (Q.J130000.2546.14H41 and QJ130000 2542 03H41) for this work. Aminu Ismaila would like to acknowledge the financial support of Ahmadu Bello University, Zaria through the TetFund intervention scheme.

References

1. N. Siu, and G. Apostolakis, A methodology for analyzing the detection and suppression of fires in nuclear power plants. Nucl. Sci. Eng. J., **94:3** (1986).
2. Berg H.P. and Hauschild J., Probabilistic Assessment of Nuclear Power Plant Protection Against External Explosions. *Nuclear Power- Practical Aspects*. (2012).
3. J. Hauschild and H.-P. Berg, How to Assess External Explosion Pressure Wave. Elect. J. of Int. Group on Rel., **01:24** (2012).
4. S. J. Jeon, B. M Jin and Y. J. Kim. Assessment of the fire resistance of a nuclear power plant subjected to a large commercial aircraft crash. Nucl. Eng. and Design, **247**:p11-22 (2012).
5. HSE, *A Review of the State-of-the-Art in Gas Explosion Modelling*, H.S.L. C. J. Lea, Buxton. p. 80. (2002).
6. A. Ismaila, R.M. Kasmani and A.T. Ramli, Consequence Assessment of Vapour Cloud Explosion Involving Large Commercial Airliner Crash upon Nuclear Reactor Containment, Chem. Eng. Trans., **56,** (2017) (to be published).
7. K.B. Mishra, K.D. Wehrstedt, and H. Krebs, Lessons learned from recent fuel storage fires. Fuel Proc. Tech., **107**: p. 166-172 (2013).
8. F. Rigas, and S. Sklavounos, Risk and consequence analyses of hazardous chemicals in marshalling yards and warehouses at Ikonio/Piraeus harbour, Greece. J. of Loss Prev. in the Process Indus., **15**(6): p. 531-544 (2002).
9. F.B.A. Beshara, Modelling of blast loading on aboveground structures—II. Internal blast and ground shock. Comp. & Stru., **51**(5): p. 597-606 (1994).
10. Baker, W.E., Cox, P. A., Westine, P. S.,. Kulesz, J. J. and Strehlow, R. A., *Explosion Hazards and Evaluation*. Elsevier (1983).
11. P.M. Locking, *The Trouble with TNT Equivalence* in *26th International Ballistics Symposium.*, Miami, FL. (2011).

12. A.C. van den Berg, The multi-energy method: A framework for vapour cloud explosion blast prediction. J. of Hazard. Mat., **12**(1): p. 1-10 (1985).
13. W.P.M. Mercx, A. C. van den Berg, C. J. Hayhurst, Developments in vapour cloud explosion blast modeling. J. of Hazardous Mater., 71(1-3) : p. 301-319 (2000).
14. J.A. Marc and E.K. Konstantinos, *Fires, Explosions, and Toxic gas Dispersions: Effects calculation and Risk Analysis*, United States: CRC Press, Taylor and Francis Group, Boca Raton,. 352 (2010).
15. A.R. Soman and G. Sundararaj, Consequence Assessment of Vapour cloud Explosion Involving Hydrogen Release. Int. J. of Emerging Tech. and Advanced Eng., **2**(11): p. 291-297 (2012).
16. IAEA, *Design of Reactor Containment Systems for Nuclear Power Plants*, International Atomic Energy Agency: Vienna (2004).
17. Lees, F. P. *Loss Prevention in the Process Industries*, 2nd Ed., Butterworth-Heinemann, Boston MA, pages 7/64 and 17/249 (1996a).
18. Netherlands Organisation of Applied Scientific Research *["TNO],"Methods for the Determination of Possible Damage to People and Objects Resulting from Releases of Hazardous Materials*", CPR 16E ("green book"), Chapter 3, "The Consequences of Explosion Effects on Humans", page 17 (1989).
19. Finney, D.I., *PROBIT Analysis*, London: Camebridge University Press, (1971).
20. IAEA, *Site Evaluation for Nuclear Installations Safety Requirement*, International Atomic Energy Agnecy: Vienna (2016).
21. V.J. Clancey, Diagonostic Features of Explosion Damage, in 6th International Meeting on Forensic Science. 1972: Edinburg, Scotland.
22. Kayes, P.J., *Manual of Industrial Hazards Assessment Techniques*. Technica. 1985, London.
23. Santamaría, J.M. and P. Braña, *Risk analysis and reduction in the chemical process industry*. 1998, London Blackie Academic & Professional.
24. J.r.m. Taveau, The Buncefield Explosion: Were the Resulting Overpressures Really Unforeseeable? Process Safety Prog., **31**(1): p. 55-71 (2012).
25. D.M. Johnson, The potential for vapour cloud explosions – Lessons from the Buncefield accident. J. of Loss Prev. in the Process Indust., **23**(6): p. 921-927 (2010).

AUTHOR INDEX

EDP Sciences – Web of Conferences
17, Avenue du Hoggar
Parc d'Activité de Courtabouf
BP 112
F-91944 Les Ulis Cedex A
France

ISBN 978-1-5108-5010-1

CURRAN ASSOCIATES INC.
proceedings.com

Thermophysical Basis of Energy Technologies (TBET-2016)

MATEC Web of Conferences Volume 92 (2017)

Tomsk, Russia
26 - 28 October 2016

edp sciences